ANDREW JACKSON

Also by Robert V. Remini

Andrew Jackson

ROBERT V. REMINI

HarperPerennial

A Division of HarperCollins*Publishers*

Library of Congress Cataloging-in-Publication Data
Remini, Robert Vincent, 1921–
    Andrew Jackson / Robert V. Remini.
      p. cm. maps
    ISBN 0-06-080132-8
    Bibliography
    1. Jackson, Andrew, 1767–1845. I. Title
E382 1969
973.5'6'0924 B—dc19   76-471

04 05 ❖ 40 39 38 37 36

For My Parents

LAURETTA AND WILLIAM REMINI

Contents

Acknowledgments

SURELY ANYONE WHO PRESUMES TO SCOOP UP A LIFE AS RICH AND exciting as Andrew Jackson's and spill it over a few hundred pages is looking for trouble. Even so it is worth a try, particularly in this period of historical writing when so many books about the Old Hero and his Democracy fall from the hands of die-hard anti-Jacksonians, euphemistically called Whigs.

Because only a fraction of the things that measure Jackson's stature and importance can be attempted here—though a careful effort has been made to touch on all the significant events in his life—I have concentrated on two aspects of Jackson's career: his skill as a politician and, to a lesser extent, his role in building the presidential office.

In the preparation of this work I am indebted to Mr. George Keenan of Fordham University for his assistance. I am also very grateful to the American Philosophical Society for a grant-in-aid to research this period of history.

Chronology

1767	Born in the Waxhaws country, South Carolina.
1780–81	Participates in American Revolution, is captured by British and later released. Receives head wound from a British officer.
1781	Learns saddlers' trade.
1784	Teaches school in the Waxhaws.
1785–87	Reads law in Salisbury, North Carolina.
1787	Admitted to the bar.
1787	Tends store at Martinsville, North Carolina.
1788	Appointed Public Prosecutor for Western District of North Carolina.
1788	Migrates to Nashville and fights first duel with Waightstill Avery.
1791	Appointed Attorney General of Mero district of the Southwest Territory.
1791	Marries Rachel Donelson Robards.
1792	Appointed Judge Advocate of Davidson County Militia.
1794	Remarries Rachel.
1795–99	Speculates in land.
1796	Delegate to Tennessee Constitutional Convention.
1796	Elected to U. S. House of Representatives.
1797	Elected to U. S. Senate.
1798	Resigns from the Senate and is appointed Judge of the Superior Court of Tennessee.
1802	Elected Major-General of Tennessee Militia.
1804	Resigns as Judge.
1805	Duels with John Sevier.
1805–7	Involved in Burr conspiracy.
1806	Duels with Charles Dickinson.
1804–12	Devotes time to plantation and other business interests.
1810	Adopts son of Severn Donelson.
1812–15	Leads troops against Indians and British.
1813	Acquires nickname, "Old Hickory."

1813	Fights with the Benton brothers.
1814	Defeats Creeks at Horseshoe Bend.
1814	Commissioned Major-General in regular army.
1815	Defeats British at the Battle of New Orleans.
1818	Leads Florida expedition.
1819	Builds Hermitage mansion.
1821	Appointed governor of Florida Territory and resigns after a few months. Resigns commission in the army.
1822	Nominated for the presidency by the Tennessee legislature.
1823	Elected to U. S. Senate.
1824–25	Wins electoral plurality but loses presidency to John Quincy Adams in the House election.
1825	Resigns from the Senate.
1825	Nominated for presidency by Tennessee legislature.
1825–28	Conducts campaign for presidency in 1828.
1828	Elected President. Rachel dies.
1829	Enunciates principle of rotation.
1829–31	Reshuffles Cabinet after Eaton affair and break with Calhoun.
1830	Vetoes Maysville Road Bill.
1830	Reopens West Indian trade.
1830	Endorses removal of Indians.
1831	Obtains most favored nation treaty with Turkey.
1831–36	Forces French to pay spoliation claims.
1832	Vetoes recharter of Second Bank of the United States.
1832	Re-elected President.
1832	Issues Proclamation over South Carolina Nullification.
1833	Orders Secretary of the Treasury to remove federal deposits from National Bank.
1834	Censured by Senate.
1834	Announces end to national debt.
1835	Escapes assassination attempt.
1836	Issues Specie Circular.
1837	Wins expunging of Senate censure.
1837	Retires to the Hermitage.
1845	Dies at the Hermitage.

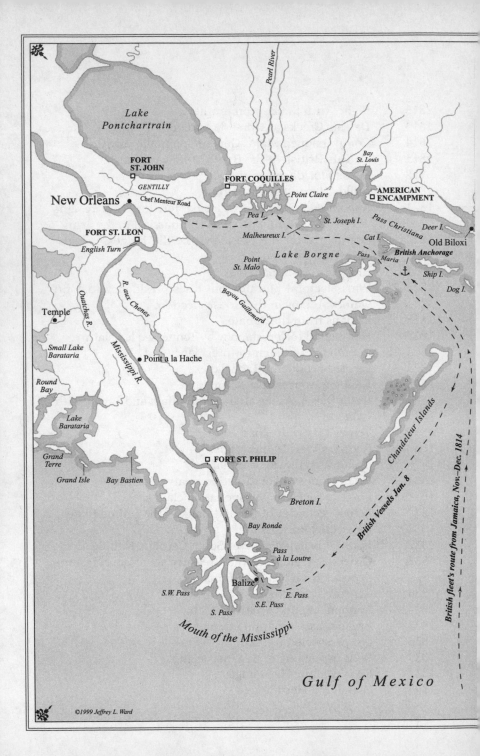

Lake Pontchartrain

FORT ST. JOHN

GENTILLY

New Orleans

Chef Menteur Road

FORT COQUILLES

Point Claire

Pea I.

St. Joseph I.

Pass Christiana

AMERICAN ENCAMPMENT

Bay St. Louis

Deer I.

Old Biloxi

FORT ST. LEON

English Turn

Malheureux I.

Cat I.

British Anchorage

Pearl River

Lake Borgne

Pass

Maria

Point St. Malo

Bayou Gaillemard

Ship I.

Dog I.

Ouatchas R.

R. aux Chenes

Mississippi R.

Temple

Point a la Hache

Small Lake Baratia

Round Bay

Lake Baratia

Grand Terre

Grand Isle

Bay Bastien

FORT ST. PHILIP

Chandeleur Islands

Breton I.

Bay Ronde

British Vessels Jan. 8

British fleet's route from Jamaica, Nov.–Dec. 1814

Pass à la Loutre

S.W. Pass

Balize

S. Pass

S.E. Pass

E. Pass

Mouth of the Mississippi

Gulf of Mexico

©1999 Jeffrey L. Ward

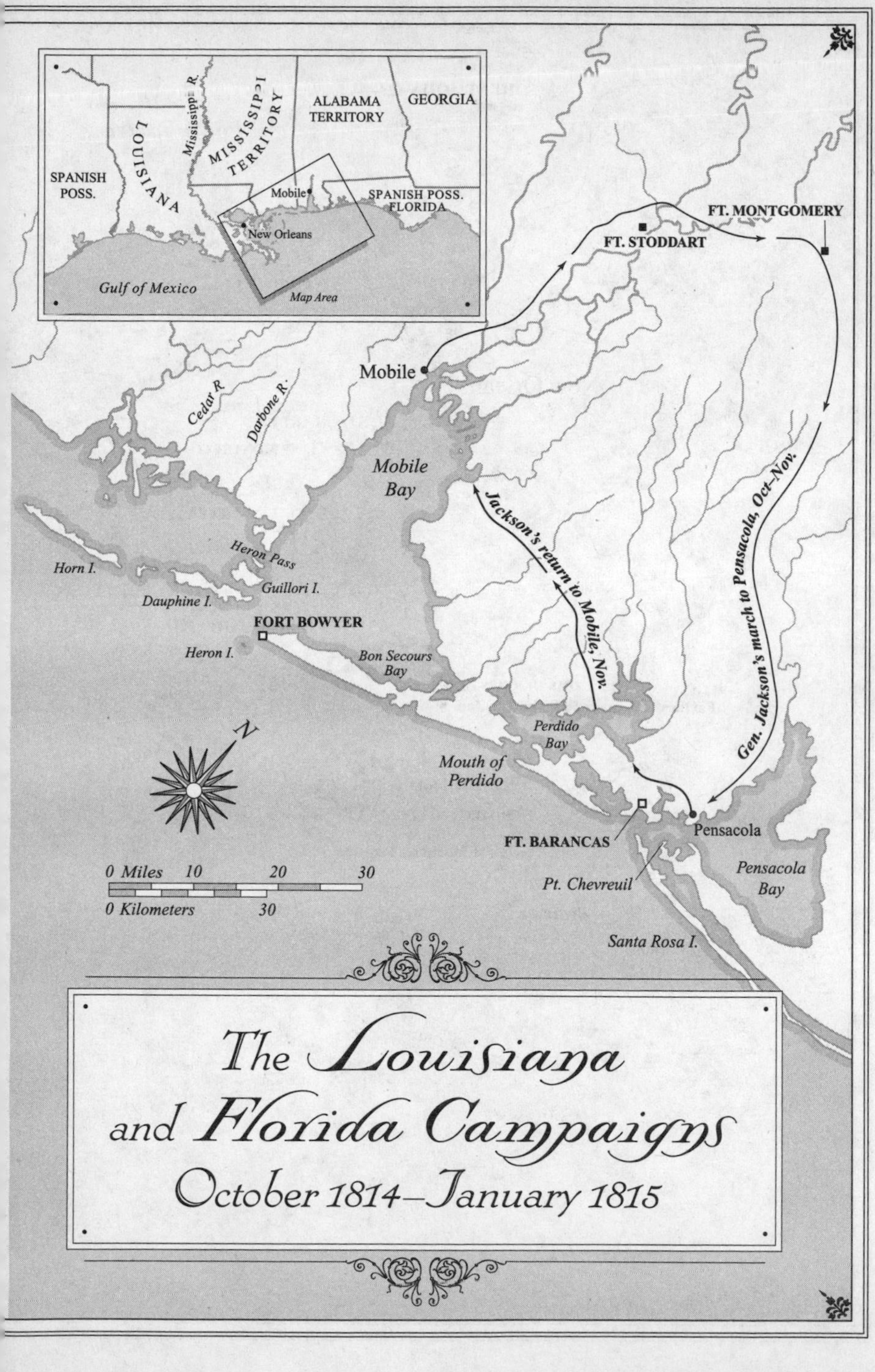

The Louisiana and Florida Campaigns

October 1814–January 1815

Inset map labels:

SPANISH POSS.

LOUISIANA

Mississippi R.

MISSISSIPPI TERRITORY

ALABAMA TERRITORY

GEORGIA

Mobile

SPANISH POSS. FLORIDA

New Orleans

Gulf of Mexico

Map Area

Main map labels:

FT. MONTGOMERY

FT. STODDART

Mobile

Cedar R.

Darbone R.

Mobile Bay

Jackson's return to Mobile, Nov.

Gen. Jackson's march to Pensacola, Oct.–Nov.

Horn I.

Heron Pass

Guillori I.

Dauphine I.

FORT BOWYER

Heron I.

Bon Secours Bay

Perdido Bay

Mouth of Perdido

FT. BARANCAS

Pensacola

Pt. Chevreuil

Pensacola Bay

Santa Rosa I.

N

0 Miles 10 20 30

0 Kilometers 30

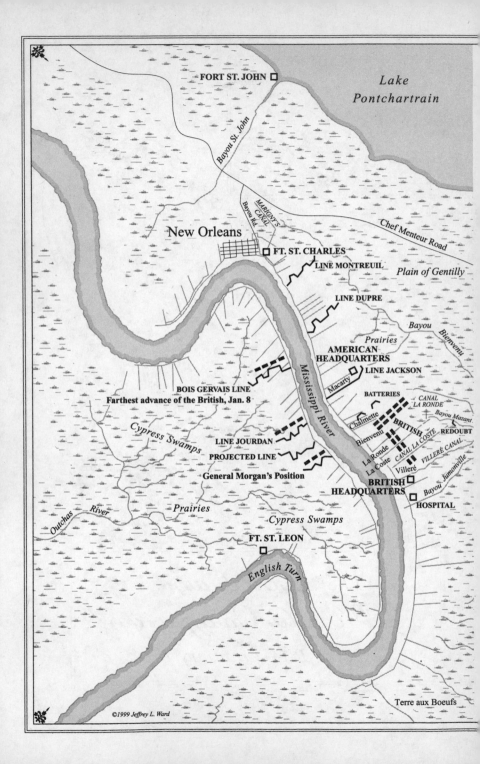

FORT ST. JOHN □

Lake Pontchartrain

Bayou St. John

New Orleans

Chef Menteur Road

MARIGNY'S CANAL
Bayou Rd.

□ FT. ST. CHARLES

LINE MONTREUIL

Plain of Gentilly

LINE DUPRE

Bayou Bienvenu

Prairies

AMERICAN HEADQUARTERS

LINE JACKSON

Macarty

BOIS GERVAIS LINE
Farthest advance of the British, Jan. 8

BATTERIES

CANAL LA RONDE

Cypress Swamps

Chalmette
Bienvenu

Bayou Mazant

BRITISH

REDOUBT

Mississippi River

LINE JOURDAN

La Ronde
La Coste

CANAL LA COSTE

VILLERÉ CANAL

PROJECTED LINE

General Morgan's Position

Villeré

Bayou Jumonville

BRITISH HEADQUARTERS

HOSPITAL □

Prairies

Outchas River

Cypress Swamps

FT. ST. LEON □

English Turn

Terre aux Boeufs

©1999 Jeffrey L. Ward

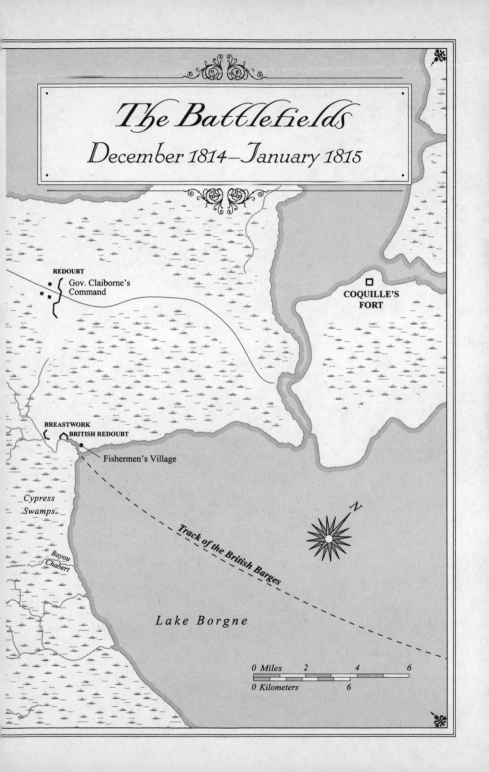

The Battlefields
December 1814–January 1815

REDOUBT
Gov. Claiborne's
Command

COQUILLE'S
FORT

BREASTWORK
BRITISH REDOUBT
Fishermen's Village

*Cypress
Swamps*

*Bayou
Chabert*

Track of the British Barges

N

Lake Borgne

0 Miles   2   4   6
0 Kilometers   6

CHAPTER I

A Roaring, Rollicking Fellow

THE ELECTION HAD BEEN AS FILTHY AS PRESIDENTIAL ELECTIONS are ever likely to get. It set a record for character assassination, scurrility, and unspeakable vulgarity. Both men seeking the high office paid a fearful price for their ambition. One of them, the sixty-year-old Andrew Jackson, watched in agony as his wife reeled under the assaults upon her character that daily spewed forth from the public prints. One day, as he sat in his home in Tennessee reading a newspaper, he spotted a paragraph that had a neatly-drawn hand pointing to the opening words. As he scanned the first line he paled; then, in a sudden, uncontrolled burst of emotion he broke down in tears, and his body shook with grief. His wife, Rachel, entered the room at that moment and, seeing his distress, asked him what was wrong. Jackson pointed to the offending newspaper. "Myself I can defend," he said. "You I can defend; but now they have assailed even the memory of my mother." Rachel picked up the paper and stared at the incredible words. "General Jackson's mother was a COMMON PROSTITUTE," it read, "brought to this country by the British soldiers! She afterward married a MULATTO MAN, with whom she had several children, of which number General JACKSON IS ONE!!!"

It may seem strange that General Andrew Jackson, hero of the War of 1812 and the courageous soldier boy of the American Revolution, could respond with tears to the lying words of a vicious newspaper editor. It would have been more characteristic had he stood up and roared his rage, summoned the vile penman to the field of honor, and there avenged his mother's name with a well-placed bullet; for

Jackson did have a monumental temper, which when roused could hurl itself with fearful fury against those who displeased him. When he chose, he could flood a room with the gorgeous sounds of Anglo-Saxon expletives. But on this occasion Jackson did not rail or rant; he wept. And this sudden change from the expected was what many of his contemporaries remembered when they later tried to catch his personality on paper and describe it to others. His was not an obvious, simple, or easy character to analyze. It was full of sharp contrasts, angular twists, and sudden turns. He was impetuous and cautious, ruthless and compassionate, suspicious and generous. He was driven by ambition—a skillful, hardheaded political operator, enamored of power, and deeply involved in all the ambiguities and oblique maneuvers that are inevitable in the pursuit of power. He was a complex of towering ambition, fierce loyalties, and stern discipline. One historian pronounced Andrew Jackson "a patriot and a traitor. He was one of the greatest of generals, and wholly ignorant of the art of war. A writer brilliant, elegant, eloquent, without being able to compose a correct sentence, or spell words of four syllables. The first of statesmen, he never devised, he never framed a measure. He was the most candid of men, and was capable of the profoundest dissimulation. A most law-defying, law-obeying citizen. A stickler for discipline, he never hesitated to disobey his superior. A democratic autocrat. An urbane savage. An atrocious saint."

But this complexity and contrast are some of the reasons for Andrew Jackson's enduring fascination. True, there were things about him that remained constant: his pursuit of fame, his conscientious performance of duty, his courage, his deep loyalties, his relentless patriotism, his everlasting harping on grievances and personal slights, and later, as President, his unshakable belief that he represented the people against aristocracy and privilege. But there was the other side. Take his celebrated temper for example; supposedly it was an uncontrolled and elemental force of nature, which when released could not be assuaged until it had run its course. Actually it was something he could turn on and off, almost at will. He frequently displayed it in bravura performances to frighten susceptible politicians. Jackson was good at this. He had a fine intuitive sense about when to scold—and also when to soothe—which in large measure explains why he made such an excellent politician and President.[1]

Occasionally, his temper did indeed race beyond his grasp. But not often; and then it was mostly when he was a young man, before his wife helped him to appreciate the value of self-control. Some said the temper was to be expected in a redheaded man who had "so much genuine Irish blood in his veins;" except that it was not Irish exactly; it was Scotch-Irish. His father and mother had come from Carrickfergus, an old town on the northeastern coast of Ireland, about ten miles from Belfast. His father, also named Andrew, was the son of a well-to-do linen weaver, and had migrated to America in 1765, with his wife, Elizabeth Hutchinson, and two sons: Hugh, who was two years old, and Robert, who was six months old. On arrival, the family headed straight for the Waxhaws, a settlement approximately 160 miles northwest of Charlestown, South Carolina, where Elizabeth's sisters were living with their husbands. This region straddled North and South Carolina and was watered by the Waxhaw Creek, a branch of the Catawba River that ran through the fertile land. The settlement was ringed by a jungle of piny woods. Near the edge of this waste, the Jacksons settled on a tract of two hundred acres, and for two years the father struggled to improve the sour land. He cleared some fields, brought in a late crop the first year, and built a cabin—to no avail. He died suddenly in February 1767 at the age of twenty-nine, leaving two boys and a pregnant wife.

A rude farm wagon ferried the body to the Waxhaw

churchyard where it was buried. Elizabeth, in no condition to return to her own house after the funeral, went to the house of her sister, Mrs. Jane Crawford, whose husband was the most prosperous of the inlaws. A few days later, the shock of her husband's death brought on labor pains, and Elizabeth gave birth to her third son on March 15, 1767. She named him Andrew after her dead husband.[2]

A controversy of sorts exists about whether Andrew Jackson was born in North or South Carolina. (It has also been suggested that he was born either abroad or at sea, but there is no validity to these theories.) The argument for North Carolina rests on the claim that Elizabeth did not go to the Crawford house after the funeral but went instead to the home of her brother-in-law, George McKemey, who lived on the North Carolina side of the Waxhaws. Jackson, himself, always believed he was born in Crawford's house, which was about a mile from the "Carolina road of the Waxhaw Creek." This would place his birth in South Carolina, and this is what most historians accept.[3]

For the first ten or twelve years of his life, Andrew was raised in the Crawford home. He attended several schools in the Waxhaws settlement, one an academy run by William Humphries and another conducted by James Stephenson. His mother, a pious Irish lady, always hoped he would someday become a clergyman of the Presbyterian Church, and she was buttressed in her hope by the fact that her young son was extremely bright and could read at a very early age. But young Andrew gave few indications that he was headed for the ministry—quite the opposite. For one thing he swore a blue streak, fine, lovely, bloodcurdling oaths that could frighten people half to death. Also, he was wild and reckless: he loved to "frolic," to dance, to play practical jokes; and best of all he loved to wrestle, jump, and run foot races. There was an air of uneasy restlessness about him, an exuberance that found outlets in outrageous tricks and games. Every now and then, he showed an ugly side that labeled him a bully. Although not a coward, he would purposely terrorize people if angered or if it suited his needs. He did this all his life, a result, no doubt, of his having been raised without a father. On one occasion as a lad, when floored by the kick of a gun he had just fired, he sprang from the ground and snarled at the boys standing around him, "By G-d, if one of you laughs, I'll kill him!" And when he spoke like that, few had the courage to call his bluff.

At Humphries' school, Andrew learned to read, write, and cast accounts. He later told a biographer that he also studied "the dead languages," by which he presumably meant Latin and Greek. Not that acquaintance with the classical languages ever particularly evinced itself in his correspondence or writings except for some legal phrases which any lawyer would know. Since he was not especially given to books he was a badly-informed man all of his life. His spelling was grotesque and his use of the English language demonstrated considerable contempt for its rules. Still, he was frequently eloquent and persuasive both in speech and writing. Indeed, when left to his own syntax, he was often powerfully expressive. Unfortunately, many of his official documents and messages as President were ruined by the turgid prose of his more educated but less inspired assistants.[4]

One story about his youth seems destined for preservation in the Jackson folklore and it rightfully needs to be "knocked in the head." This is the story that at the age of nine or ten he was called upon by the community to read aloud the Declaration of Independence, recently adopted by the Second Continental Congress at Philadelphia. It is a charming story and deserves to be true; unfortunately it is the fiction of a twentieth century writer who made it up out of whole cloth to add color and dimension to his account of Jackson's early life.[5]

There was no need to embellish the record. Enough factual

material exists about Jackson's real life to tell several exciting stories. Indeed, once the American Revolution began, traces of the heroic become apparent in all the members of the Jackson family. This was not surprising since the Revolution nurtured something already planted by the mother: a ferocious hatred for the British. During many winter evenings, Mrs. Jackson would gather her sons around her and chill them with stories about the terrible sufferings their grandfather endured at the siege of Carrickfergus and about the brutal treatment the British inflicted on the laboring poor. The children trembled at the horror of her stories. The Jackson boys developed an abiding hatred for the oppressors of their people so it was not surprising that with the outbreak of the American Revolution the oldest Jackson boy, Hugh, who was still in his early teens, joined the regiment of William Richardson Davie to strike a blow for Ireland, his grandfather, and freedom. He fought at the Battle of Stono Ferry and died shortly thereafter, not from wounds, but from the "excessive heat of the weather, and the fatigues of the day."

The fighting and misery of the Revolutionary conflict came to the Waxhaws late in the spring of 1780 when British commander Lieutenant-Colonel Banastre Tarleton with a force of three hundred horsemen surprised a detachment of patriot soldiers, killing 113 and wounding 150. Because many bodies were found broken and mangled, with a dozen or more wounds inflicted on some of them, the engagement was called a massacre, and Tarleton a butcher. Survivors were taken to the Waxhaws meeting house, which had been converted into a hospital, and were tended by Mrs. Jackson and her two sons.

All that summer there were alarms of troop movements of the "murderous Tories." Andrew and his brother, Robert, accompanied the patriots on several expeditions, and were present at the attack on the British post of Hanging Rock, where the Americans would surely have gained a victory if they had not paused to drink captured rum. On this expedition the Jackson boys rode with Colonel Davie. Andrew was thirteen, and while it is unlikely he did more than just attend the troops and carry messages, he may have watched and studied Davie as the commander conducted the campaign. Like the future General Jackson, Davie was bold in planning his moves, yet cautious in their execution. He was "untiringly active; one of those cool, quick men who apply mother-wit to the art of war. . . ." He spent sleepless nights in vigil; he paid close attention to details; and when he moved his troops he did it swiftly. Andrew greatly admired Davie, and it was the belief of an early biographer that, insofar as any man served as Jackson's model for soldiering, that man was William Richardson Davie.[6]

On the sixteenth of August, 1780, American patriots in the South sustained a crushing defeat at Camden by General Charles Cornwallis. Following the victory, Cornwallis turned his army toward the Waxhaws. For the next year, the Carolina countryside was one vast charnel house of butchered Tories and patriots. It was no longer a revolution, but a civil war, with brother fighting brother, father against son, and neighbor killing neighbor. In later years, Jackson remembered the fury of those days, especially the case of one man who, after finding a friend murdered and mutilated, dedicated himself to slaying Tories. This demented man scoured the countryside, hunting for victims; he lay in wait for them; he shot them on sight; and before the war ended he had quenched his vengeance by killing twenty loyalists. When the madness finally subsided he was a conscience stricken wretch who lived out his days with the horror of his deeds constantly before him.

Following one minor engagement, in which British dragoons sent Americans scurrying, Andrew and his brother took refuge in the house of Thomas Crawford. A Tory neighbor informed the British of their whereabouts. The house was soon surrounded, and the boys were taken prisoner. The soldiers began to pillage the house, smashing furniture, breaking glasses, and tearing clothes into rags. While this was in progress the officer in command of the party ordered Andrew to clean his jackboots which were encrusted with mud. Andrew refused. According to one biographer, his refusal was flung out with the words, "Sir, I am a prisoner of war, and claim to be treated as such." Whereupon the officer lifted his sword and aimed it straight at Andrew's head. Instinctively, the young man ducked and raised his left hand in time to break the full force of the blow. He received a deep gash on his head, however, and another on his hand—two souvenirs of British sentiment he took with him through life. The officer then ordered Robert to clean the boots. Again a refusal, and this time the officer hit Robert so hard with his sword that he sent him sprawling across the room.

The two brothers, along with twenty other prisoners, were later taken on horseback to Camden, a distance of forty miles. It was a long and painful journey for the wounded boys, made worse by the lack of food and water. But this only prepared them for the agony of Camden. There, they were huddled into an enclosure with 250 other prisoners, cursed for being rebels, and assured they would soon be hanged. No beds were furnished; no medicine; little food, only a small amount of bread. When the British discovered Robert and Andrew were brothers, the boys were quickly separated. Their shoes and jackets were stolen; they sickened and turned yellow; and Andrew, "chafing with suppressed fury . . . passed now some of the most wretched days of his life." Soon, smallpox broke out among the prisoners, killing about a tenth, and disfiguring most of the others. Both brothers caught the dread disease, and both might have succumbed in the prison had their mother not arrived at the time when an exchange of prisoners was being arranged. Thirteen British soldiers were eventually surrendered in return for the

Jackson boys and five of their neighbors. Mrs. Jackson procured two horses, placing the dying Robert on one, and riding on the other herself. Poor Andrew had to walk the forty miles back home. Wearily, "he dragged his weak . . . limbs, bare-headed, barefooted, without a jacket; his only two garments torn and dirty." On the final leg of the journey, a cold, drenching rain lashed the trio. When they arrived home, the boys were put to bed. "In two days Robert Jackson was a corpse, and his brother Andrew a raving maniac."[7]

His mother's nursing skill and the attention of a Dr. Tongue brought about Andrew's recovery, but it took several weeks of tireless care and devotion. Many months followed before he could be left to himself. With her son out of danger, this incredible woman set off with two other ladies from the Waxhaws to travel the 160 miles to Charleston, there to nurse the prisoners of war held in prison ships, some of whom were Mrs. Jackson's kinsmen. A tradition exists that these heroic ladies traveled on foot, but more probably they rode on horseback. They did what they could to bring comfort to the men confined to these floating prisons of disease and death. A short time later, while visiting a relative, William Barton, who lived near Charleston, Elizabeth Jackson was taken ill with the cholera, or "ship fever" as it was called, and died after a short illness. She was buried in an unmarked grave. A small bundle of her possessions was sent to her son at the Waxhaws as a kind of formal notice of her death.

Elizabeth Jackson was an extraordinary woman of courage, high purpose, and fantastic interior strength. If her life and death taught her son anything at all, it taught him something about the meaning of personal dedication and responsibility. And no one has ever seriously questioned Jackson's courage or his sense of duty. Those who knew his family background understood a little about where and how he had obtained them.

So, at the age of fourteen, Andrew was orphaned by the war. One way or another the British had robbed him of mother and brothers and had left him with two visible scars to serve as a perpetual reminder of their regard. Orphaned though he was, he still had family—uncles, aunts, and cousins—and so with little thought he went to live with his uncle, Thomas Crawford. Unfortunately, it did not work out. He was sullen, depressed, and angry. Something of his mood can be glimpsed by what happened shortly after arriving at his uncle's house. A soldier, one Captain Galbraith, was also living there and for an unknown reason objected to something Andrew said or did. As the captain raised his hand to chastise him, Andrew exploded in a rage and swore that if he was struck, Galbraith was a dead man. The hand drooped. But there was such bad feeling between them that Andrew packed his belongings and went to live at the house of another relative, Joseph White. This proved to be a better arrangement, for White's son was a saddler, and Andrew loved horses and everything connected with them. He worked in the saddler's shop for six months and learned the rudiments of the trade. It was a brief experience but one he thoroughly enjoyed.[8]

In 1781, the same year Mrs. Jackson died, Cornwallis surrendered at Yorktown. Although Charleston continued to be occupied by the British for another fourteen months, the war was all but officially over. Several of Charleston's more socially prominent families waited for the British evacuation from the safe distance of the Waxhaws settlement, and while they waited Andrew took up with their sons and for a time led a wild and merry life: gambling, drinking, cockfighting and horse racing—mostly horse racing, a sport he could never resist and on which he lost more money than he could afford. After the British left, he went to Charleston to claim a three or four hundred pound inheritance left him by his grandfather in Carrickfergus. He dissipated this legacy in one glorious spree. He was headed for financial trouble if not prison when he came upon a dice game called "Rattle and Snap" and was challenged to try his luck. A player staked $200 against Andrew's horse. Andrew quickly dispelled his doubts and accepted the challenge. He picked up the dice, warmed them with his will and his need to win, and then let them fly to his salvation. He won the wager, paid his debts, and rode triumphantly out of Charleston.

After that brush with near-disaster Andrew decided he must reform his life. In a moment of excessive repentance he determined to become a schoolteacher! His own education needed attention, but that did not trouble him. Once he made up his mind there was no stopping him. So for a year, possibly two, Andrew taught school in the Waxhaws country, and he diligently absorbed great chunks of knowledge the day before he passed them along to his students. However, for a fun-loving seventeen-year-old youngster, schoolteaching could not hold his attention for long. Besides, there was little money in it, and the profession could never support his horse racing and cockfighting fancies. More important, as a shrewd and perceptive young man, he soon realized that teaching held few prospects for the future. Now that the peace treaty with Great Britain had been signed ending the Revolution and a new nation was launched, it was the legal profession that would shape the future direction of the country. Most of the old Tory lawyers were gone; there were new problems and new opportunities that "threw into the hands of the whig lawyers" a "lucrative business." For a "young man on the make" the legal profession held much promise, and Andrew was both young and "on the make."

In December, 1784, Andrew gathered his money and belongings and rode north to Salisbury, North Carolina, a distance of seventy-five miles from the Waxhaws, where he entered the law office of Spruce McCay to learn what he had decided would be his new profession. Salisbury was the seat of Rowan County and was situated in the midst of the red clay section of North Carolina. At the town's modest tavern, called the Rowan House, Andrew found quarters and the kind of company that cheered his exuberant spirits, especially after a long day of reading law at McCay's office. For the next two years, he diligently copied papers, ran errands, cleaned the rooms, and did all the other necessary things that supposedly prepared him for admission to the bar. Then, when the business of the day ended, Andrew took off with several other students, one of whom was John McNairy, and together they "burned up" the town. He was really good at this, achieving the rare distinction of being named the leader of all the rowdies and misfits in the community. Said one: "Andrew Jackson was the most roaring, rollicking, game-cocking, horse-racing, card-playing, mischievous fellow, that ever lived in Salisbury." Said another: "He did not trouble with the law-books much; he was more in the stable than in the office."[9]

There was a dancing school in Salisbury, and naturally the rip-roaring Jackson attended it regularly. Probably, it was here that he learned many of the social graces that observers noticed about him later on. Anyway for some unaccountable reason, the school, in a moment of madness, appointed him to serve as a manager of the Christmas ball. As a joke, just "to see what would come of it," Jackson sent invitations to Molly Wood and her daughter, Rachel, the town's fanciest prostitutes. But the women, instead of realizing it was all a joke, went to the ball. When they sauntered in unannounced, decked out in all the colors of the rainbow, the dancers jolted to a stop. The respectable young ladies withdrew to one side of the room, giggling and pointing at the overdressed doxies. Forthwith, the two cruelly abused women, their invitations visible in their hands, were ushered from the room, and the dance began again. Jackson was tongue-lashed for his impertinence, after which he apologized profusely, explaining that it had all been meant as a joke and that of course he would never have issued the invitation had he believed that the sporting ladies would misunderstand and accept.

On another occasion, Jackson and his buddies were celebrating in the tavern. It turned out to be a wildly delightful night of pure, distilled joy. So successful was it that all agreed that the glasses hallowed by such a magnificent affair should never be desecrated by future use. They were all smashed. And if the glasses were sacred, how about the table on which they had rested? Away went the table—broken beyond repair. Next, the chairs joined the litter; then the bed. Following these, the clothes and curtains that had observed the sacred rites were shredded and piled into a heap, and set afire. On and on they went until the room was a total, burned out shambles. Oh, it was a perfect night, a night to remember—forever.

Many years later, in 1824, the good people of Salisbury— particularly the ladies who remembered the Christmas ball—were appalled by the news that Jackson was running for the presidency. "What!" cried one. "Jackson up for the President? *Jackson? Andrew* Jackson? The Jackson that used to live in Salisbury? Why, when he was here, he was such a rake that my husband would not bring him into the house! It is true, he *might* have taken him out to the stable to weight horses for a race, and might drink a glass of whiskey with him *there*. Well, if Andrew Jackson can be President, anybody can!"

Still, with all his roaring and rollicking, he learned the law, or at least that part of it that suited his needs. And what Jackson needed, he worked at, however much he enlivened it with card-playing, drunken parties, cockfighting and such pranks as moving outhouses to remote fields—one of his less accommodating practical jokes. Years later, someone from Salisbury visited him in the White House and asked about his stay in the town. Jackson blushed, and then his eyes twinkled a little. "Yes, I lived at old Salisbury," he said. "I was but a raw lad then, but I did my best."[10]

Early in 1787, Andrew moved from Salisbury. Considering his prodigality, it is quite possible the people of the town made life so uncomfortable for him that he had to leave. Whatever the reason, he left McCay's establishment and spent six months at the office of Colonel John Stokes, one of the most brilliant men of the North Carolina bar. Stokes had lost a hand at Buford's Defeat. In its place, he wore a silver knob that he often banged down hard on a table to emphasize a point. Under the tutelage of this learned and slightly eccentric man, Jackson completed his legal training, and on September 26, 1787 appeared for examination before Judges Samuel Ashe and John F. Williams. Since they found him a man of "unblemished" moral character and one competent in the law they declared that he was authorized to practice as an attorney in the county courts of pleas and quarter sessions within the state of North Carolina.

But, as Jackson soon discovered, there was little law business for a young man of twenty, who was just beginning his profession. For the next year, he poked around Martinsville in Guilford County, North Carolina, staying with two friends who kept a store. He helped out in the establishment and in that way learned still another trade. He was good for business because people liked him and felt comfortable with him. Most agreed that he possessed that indefinable quality called presence. Wherever he was, whatever he was doing, it seemed the most natural thing in the world for him to direct what was going on. He was not always good at what he directed—he had apprenticed too many trades to have mastered any one of them—but whenever he appeared, there was never a doubt who was in charge. His very appearance commanded attention. Straight and tall, he stood six feet one inch in his stockings. His face was long and thin, and his hair, reddish in color, was quite abundant and fell over his forehead, hiding in part the scar awarded him in the Revolution. The most striking characteristic of his face was his deep blue eyes—eyes that could blaze with such passion when he was aroused as to paralyze with fright those who were the objects of his wrath. And during his youth, a word, ill-conceived or thoughtlessly thrown out, was enough to set his eyes blazing. Yet, those who stood near him at the moments when he erupted and watched as he screamed and foamed at the mouth noticed that there was something very strange indeed. He appeared to be going through the motions of a great passion without being involved; even during the height of the screaming and raving, he appeared to be in absolute control of himself. The whole scene seemed fake, acted out deliberately for some predetermined reason. *"No man,"* wrote one, *"knew better than Andrew Jackson when to get into a passion and when not."* All of which suggests that there was more art—and deceit—in Jackson than most suspected. And more self-discipline, too. Granted that he did some wild and foolish things in his time, yet he rarely attempted what his own quick-thinking brain warned him was beyond his reach. He was extremely calculating, and although he could be bold in the conception of his plans, he was invariably cautious in carrying them out.

These, then, are some of the bits and pieces that went into fashioning Jackson as a man and politician. And in 1788, politics was not very far from his thoughts. He had learned that his old friend, gaming companion, and fellow student of his Salisbury days, John McNairy, had been elected by the North Carolina legislature to be the Superior Court judge for the western district of the state, which stretched westward to the Mississippi River and had little colonies of settlers huddled along its several rivers. As judge, McNairy had authority to appoint the public prosecutor for the district, and Jackson now decided he wanted this job and prevailed on McNairy to give it to him. Another friend, Thomas Searcy, solicited the office of clerk of the court, and three or four other comrades, all lawyers, took it into their heads to go along too, thus rounding out a small band of young, energetic men who saw, or thought they saw, their future and fortune in the West.

The country they planned to enter had a troubled history. It lay west of the Alleghenies and was undergoing settlement in two general areas: the eastern section around the Watauga and Nollichucky rivers and extending as far as the neighborhood of Knoxville; and the western section located in the Cumberland River Valley. At one point in the turbulent history of these counties, the settlers struck for independence. They created a new state, called it "Franklin," and elected John Sevier as governor. But North Carolina regarded their presumption as an act of treason, and after considerable persuasion convinced the westerners to return to their former allegiance.

It was at this juncture in the history of North Carolina that Jackson, McNairy, and their friends were about to cross the mountains and head for Nashville. Early in the spring of 1788 they rendezvoused at Morganton, North Carolina. Each man was equipped with a horse, a few belongings, a gun, and a wallet containing letters from distinguished citizens of the old community to the settlers of the new. And off they went along the mountain trace, this unlikely crew of judge, prosecuting attorney, clerk, and lawyers. If nothing else, they would bring a massive dose of law to the West.[11]

CHAPTER II

Tennessee Lawyer

THE COMPANY WAS ASLEEP—ALL BUT ANDREW JACKSON, WHO SAT on the ground, his back against a tree, silently smoking a corncob pipe. At about ten o'clock, as he began to doze, he heard the hooting of owls far off in the depths of the forest. Strange to hear owls in this country, he thought. Owls! Instantly, Jackson bolted to his feet, his hand scooping up his rifle as he raced to his friends. "Searcy," hissed Jackson, as he shook his friend to consciousness, "raise your head and make no noise."

"What's the matter?" responded the irritated Searcy.

"The owls," said Jackson, "listen—there—there again. Isn't that a little *too* natural?"

"Do you think so?" stammered the now awakened and thoroughly frightened Searcy.

"I know it," replied Jackson. "There are Indians all around us. I have heard them in every direction. They mean to attack before daybreak."

The other men were hurriedly awakened. Jackson told them what he knew and advised that they immediately break camp and start running as fast as they could. All agreed, and with the haste born of fear they fled the camp and plunged deeper into the forest. Fortunately, they did not see or hear any Indians for the remainder of the night. However, several hours later, a party of hunters came upon the abandoned site and decided it was a good place to spend the night. When they realized what the hooting meant it was too late. The Indians killed all but one of their party.

That was the only frightening incident of the trip to the

West, although something else happened along the way that could have had tragic consequences. It occurred at Jonesborough, where the party stayed for a few months to attend a session of the district court. Adjusting quickly to his new surroundings, Jackson swung into the rhythm of the town by personally riding in a horse race, buying a Negro girl, and being admitted to practice in the local court. In one case tried before this Jonesborough court, Jackson bumped into Waightstill Avery, a local lawyer, who resorted to sarcasm to rebut the arguments of his young opponent, and the trigger-tempered Jackson, in a rage, tore out the leaf of a law book, scribbled a few lines on it, and hurled it at Avery. What he wrote is not known—something provocative no doubt, but it was by no means a formal challenge. He needed time to think about whether he wished to risk his life over this trifle, and only after spending the night in contemplation did he finally decide to demand satisfaction. "You recd. a few lines from me yesterday and undoubtedly you understand me," he wrote to his adversary. "My charector you have injured; and further you have Insulted me in the presence of a court and a larg audianc. I therefore call upon you as a gentleman to give me satisfaction for the Same; . . . and I hope you can do without dinner untill the business is done; for it is consistent with the charector of a gentleman when he Injures a man to make a spedy reparation; therefore I hope you will not fail in meeting me this day."

Avery, a much older man, was no duelist, and he disapproved the code; but in the western country he had no choice if he expected to stay in business; so, reluctantly, he met Jackson in a hollow north of the town just a little after sundown. The distance was measured and the duelists took their positions. A signal was given, and both fired simultaneously. Fortunately, both missed. His honor now restored, Jackson strode up to his adversary, announced he had nothing further to settle, shook hands with Avery, and walked away.

This was Jackson's first known duel. Since he was not an especially good shot and continued to get into all sorts of scrapes later on, he had fantastic luck in escaping with his life. But some of his adversaries came very close—very close indeed.

A few months later, Jackson and his party left Jonesborough and headed for Nashville, arriving there on October 26, 1788. As they neared the town, they paused on the bluff overlooking the settlement and saw stretched out before them a vast fertile country, watered by the great Cumberland River. Years earlier, there had come into this lush paradise a small group of explorers, commanded by Captain James Robertson. They were followed by Colonel John Donelson who guided the families of the explorers and brought them inland by a treacherous voyage down the Holston River to the Tennessee, then down the Tennessee to the Ohio, up the Ohio to the Cumberland, and finally along the Cumberland to Robertson's settlement. The colonists numbered about 120 men, women, and children. They were a tough and rugged breed, among whom were Donelson's eleven children, the youngest being a thirteen-year-old girl, named Rachel.[1]

This beautiful and gently undulating country, well-watered and well-stocked with animals and sturdy trees, would ordinarily not be wrested from the Indians without a fierce struggle. Fortunately for the settlers they infected the savages with smallpox as they rode down the rivers to Nashville, and hundreds of Indian braves died before they could seriously challenge the invaders. Although the white men successfully planted their colony, it remained hazardous for many years, with the rigors of the wilderness and the constant menace of the Indians taking a heavy toll.

Eight years after the settlement was planted, Jackson and

the other members of the legal aid society arrived. By that time, the Nashville community consisted of a court house, two stores, two taverns, a distillery, and a number of cabins, tents, houses, and other nondescript shelters. By that time, too, Donelson was dead—murdered either by a white robber, which was what the family believed, or by an Indian. His youngest daughter, Rachel, had grown into a dark-eyed, dark-haired, vivacious woman. She married Lewis Robards, a Kentuckian of good family who loved his wife but within a short time grew neurotically suspicious of her coquettish ways and began believing all sorts of improprieties about her. Indeed when Jackson arrived at Nashville, Rachel was separated from her husband, having been ordered out of his house in Kentucky for talking to another man in a manner which Robards felt exceeded ordinary politeness. A little later, after trying unsuccessfully to live without his wife, Robards admitted his mistake and asked forgiveness. Rachel gave him a second chance and the reunited couple moved in with her widowed mother at Nashville until such time as the Indians had been sufficiently subdued to permit building their own house beyond the settlement.

At length the fun-loving Jackson rode into Nashville. The widow Donelson took him in as a boarder because of the additional protection he could provide the large family against marauding Indians. The house was bigger than any other in the area and appeared comfortable, so Jackson agreed to join the family. Moreover, the Donelsons were one of the first families of Nashville and Jackson, after one icy, appraising look, thought the association could certainly do his career no harm. But a worse arrangement could hardly be imagined. Here was Rachel, a high-flying, lively and delightful young girl, who liked to dance, ride horses, and tell amusing stories to an appreciative audience; and here was Jackson, who could match Rachel's gaiety and fun and maybe spice it with a little wildness; and here, too, was

Robards, almost pathologically suspicious of his wife's behavior—all living under the same roof. A violent misunderstanding, if not something worse, seemed imminent.

For a time, all was serene. Jackson plunged into his law duties and rapidly drilled though a mountain of work that awaited him as prosecuting attorney. Debtors had refused to pay their legitimate obligations and had banded together to defy the law. They had working for them the only licensed attorney in the district, and since the sheriff was incompetent, creditors despaired of obtaining justice. Then the new prosecutor arrived—and suddenly the propertied classes in Nashville found their long-awaited savior. Within a month, he had enforced seventy writs against delinquent debtors. Naturally, their reversal infuriated the debtors, and one of them walked up to Jackson one day and to show his displeasure about the way things were going deliberately stepped on the young man's foot. Without batting an eye, Jackson turned around, picked up a piece of wood, and calmly knocked the man out cold. Respect for the law, Jackson-style, had arrived in Nashville.

Immediately, creditors were banging at Jackson's door begging him to handle all their business—to prosecute this defaulter, go after that one, draw up contracts, and check into land titles. So promptly and expeditously did Jackson administer the business they thrust at him (and he relied less frequently on a club or other weapons) that within a relatively short time he had firmly "secured his career at the bar of Tennessee." For the next ten years, he practiced law throughout a large part of the then southwestern district of the United States, following the circuit courts and shuttling back and forth from Nashville to Jonesborough or Gallatin, or some other town where a court was sitting or a case might take him. Most of his cases dealt with land titles, debts, assault and battery, and sales. His practice expanded so rapidly and he was so successful and happy at it that he decided to remain permanently in Nashville. The records of Davidson County, of which Nashville was the seat, show that of all the cases argued in the county during the first years after his arrival, Jackson handled between a fourth to one half of them. Obviously he had more cases than he could properly manage but financially he made out quite well. Since money was scarce in the West, he accepted land in payment for his services, and quickly became one of the most important landowners in Tennessee.

In 1791, Jackson was appointed attorney-general for the Mero district of the Southwest Territory and a year later became judge advocate of the Davidson County Militia. For these offices he could thank William Blount, the newly appointed territorial governor. The Southwest Territory had been created by an act of Congress in 1790 and extended from Kentucky to the present states of Alabama and Mississippi. In 1791, the northern portion of this territory, Kentucky, was admitted as a state in the Union. Blount continued as governor over the remaining territory and he named Jackson as attorney-general of the Mero district with duties similar to those he previously had as prosecuting attorney. Thus, almost from the beginning of his career in the West, Jackson accepted favors and was identified with the Blount faction in Tennessee politics. The party that developed in opposition to Blount clustered around the person of John Sevier, the territory's popular hero who had won a great victory at the Battle of King's Mountain during the Revolution.[2]

However, while the young solicitor's legal career prospered, his private life became more entangled. Robards, not a little justified, had grown increasingly resentful of Jackson's presence in the Donelson home and at one point became so aggravated and suspicious that he threatened to whip Jackson if he were provoked any further. Demonstrating caution and good sense, which he could summon when necessary, Jackson left the house shortly afterward and went to live at Casper Mansker's station, while Robards returned to Kentucky without his wife. But again, the unhappy husband found life unbearable without her, won forgiveness a second time and succeeded in talking Rachel into returning with him to Kentucky. Yet no sooner were they back at their home than the quarrels started again, whereupon Rachel summoned her family to come and get her and take her back to Tennessee. However, none of her many brothers responded to her call. Who should poke his head through the door of the Robards' homestead to do the escorting honors but the daring and intrepid Andrew Jackson. And this surely was absolute folly or absolute calculation. His appearance confirmed Robards' worst suspicions, and it involved Jackson in an overt act that was certain to injure both Rachel's and his own reputation. Why, then, did he do it? Why did he look for trouble? Undoubtedly by this time, he was deeply in love with Rachel. Still his dash to Kentucky was not the action of an impetuous lover. Probably, he was asked to go by the Donelson brothers—who were themselves pretty disgusted with the affair—and in accepting, he deliberately laid the grounds for a divorce. Despite the inevitable scandal, he had decided to marry Rachel because such an alliance with one of the most important families in the territory was precisely what he wanted. This is not to imply that he did not love Rachel and love her very deeply. There can be no doubts on that score. It is merely to suggest that he carefully weighed the pros and cons of the marriage and the consequences of his action, and only then did he rush off to Kentucky to get her. For the rest of his life, Jackson remained sensitive to the charge that he prevailed upon Rachel to desert her husband. He was sensitive to the charge because, to a large extent, it was true.

Nor was Rachel entirely innocent. (In her old age she

became excessively pious which could be interpreted as the sign of a guilty conscience.) Even if she did not betray Robards, her actions during the years in Nashville—dancing, flirting, and the rest—were the actions of a girl being courted, not those of a properly married young woman.

In any event, Jackson escorted Rachel back from Kentucky to the home of her sister, Jane, the wife of Colonel Robert Hays, the commander of the Mero cavalry regiment. In the fall of 1790, a report circulated that Robards planned to return to Nashville and take his wife back to Kentucky with him. Rachel was so frightened at the prospect of confronting her husband and so anxious to give him additional evidence for a divorce that she fled to Natchez. Colonel Stark, a family friend, agreed to accompany her. Since it was a perilous voyage and liable to Indian attack, old-dependable Jackson was again prevailed upon to see the lady safely away from the clutches of her husband. The Donelson brothers, of whom there were many and who could easily have obliged, again stayed home. Although "bowed down with remorse" for "having innocently and unintentionally been the cause of the loss of peace and happiness of Mrs. Robards," Jackson accompanied her down the Cumberland and Mississippi rivers. He dutifully delivered her to friends, after which he returned overland to Nashville, going by way of the extremely dangerous Natchez Trace. Not much later, he heard that Robards had been granted a divorce in Virginia, obtained through the good offices of Major John Jouett, Robards' brother-in-law and a member of the state legislature. Without bothering to check this hearsay evidence, lawyer-Jackson sped back to Natchez and in August, 1791, married the woman of his choice.

Actually, Robards had no divorce; what he had was an enabling act permitting him to bring suit against his wife in a court of law. The act stated that if a jury found Rachel guilty of adultery and desertion as charged, then a divorce would be granted. For some reason Robards did not immediately press the action in the courts. Possibly, he hoped for a reconciliation which, considering the number of times he had already asked Rachel to take him back, was not improbable.

Slightly more than two years later the Jacksons learned the awful truth, that not until September 27, 1793 had Robards received a divorce from the Court of Quarter Sessions in Mercer County, Kentucky. The news shattered the couple. There can be no doubt that both sincerely believed they had been married in Natchez, but in the eyes of the law their relationship was illegal. John Overton, a friend and fellow boarder at the Donelson home, suggested they rewed. Jackson at first refused, arguing that everyone in the territory knew they were properly married. But the wisdom of the Overton suggestion finally convinced him and on January 17, 1794 he and Rachel recited their wedding vows a second time.

Whatever the unusual and unfortunate circumstances of their marriage, the life Andrew and Rachel spent together thereafter was indeed happy, although marred periodically by intruders whose tongues darted with slander. The quarrels, the bitterness, and the accusing words that typified Rachel's marriage to Robards were absent from her marriage to Jackson. The love they bore each other matured and deepened over the years. Yet, though there was great warmth and affection between them, and though they lived together for over thirty years, they were outwardly very formal and proper with each other, as was the custom in those days. Even in the matter of address they were formal. He called her "Mrs. Jackson" or "wife"—never "Rachel"; while she addressed him as "Mr. Jackson,"—not "General" and not "Andrew." And no one ever called him "Andy!"[3]

Despite the divorce and its implications the marriage was very advantageous for Jackson. It linked him to one of the first and largest families in Tennessee and automatically gave him social position. Indeed, within days after his return to Nashville following the initial wedding, Jackson was elected a trustee of the Davidson Academy, a group of the "first" men and clergymen in Nashville. But he had expected advantages from his marriage, for as he later explained to his ward, Andrew J. Hutchins, a man must use his head in selecting a wife. "One word to you as to matrimony," he wrote, "—seek a wife, one who will aid you in your exertions in making a competency and will take care of it when made, for you will find it easier to spend two thousand dollars than to make five hundred. Look at the economy of the mother and if you find it in her you will find it in the daughter. recollect the industry of your dear aunt [Rachel], and with what economy she watched over what I made, and how we waded thro the vast expence of the mass of company we had. nothing but her care and industry, with good economy could have saved me from ruin, if she had been extravagant the property would have vanished and poverty and want would have been our doom. Think of this before you attempt to select a wife."

After the happy couple returned from Natchez—they were both twenty-four years old at the time—Jackson bought a small plantation for his home, named Poplar Grove, which stood on a hairpin turn of the Cumberland River. As he had anticipated, Rachel proved indispensable to him in running the plantation, and he later acknowledged that without her their estate would have dwindled to a pittance. Since he was away much of the time, even before his military and political careers began, her managerial skill was repeatedly tested. In fact, it now seems certain that the success of the plantation was due in large measure to Rachel's extraordinary administrative talents. Perhaps she, rather than Jackson, was the real "businessman" in the family.[4]

As his law practice continued to prosper, the attorney-general bought a larger plantation in 1795, called Hunter's Hill, and for the next two years "laid the foundations of a large estate," his name appearing frequently as the purchaser and assignee of many sections of land. At length, he bought a tract of 650 acres for $800 that formed the basis of his plantation, Hermitage. This was a square mile of fertile and rolling land, approximately twelve miles from Nashville. Thus, by the time Tennessee entered the Union as a state in 1796, "Jackson was a very extensive landowner," moderately well-to-do and, socially, a member of the upper class. "The secret of his prosperity," wrote one, "was that he acquired large tracts when large tracts could be bought for a horse or a cow bell, and held them till the torrent of emigration made them valuable."

This "torrent of emigration" crested so rapidly that the territorial legislature directed that a census be taken to determine whether there were sixty thousand inhabitants, the minimum required for admission to statehood. If there were sufficient people in the area, the territorial governor could then call a constitutional convention and begin the process of state-making. The census completed, it was learned that 77,262 inhabitants lived in the territory, of whom 10,613 were slaves. Without delay, the governor ordered the election of delegates—five from each county, for a total of fifty-five—to assemble on January 11, 1796 at Knoxville to draft a constitution. Davidson County elected five of its very best men. They selected James Robertson, John McNairy, Thomas Hardeman, Joel Lewis, and Andrew Jackson.

The convention met in a small building in Knoxville on the outskirts of town. The munificent sum of $12.62 was appropriated to outfit the hall: $10.00 for seats, and the rest for an oilcloth to cover the chairman's table. The legislature had allowed each member $2.50 per diem but had forgotten to appropriate funds for the secretary, printer, and doorkeeper. In a unique display of generosity and harmony, the members rectified this oversight by changing their own fee to $1.50 and directing the difference be given to the unprovided officers. That done, the delegates proceeded to business and completed their task in twenty-seven days; what is more, they wrote a workable democratic document. According to the adopted constitution, any white man could vote after a residence in the state of six months—a freeholder could vote as soon as he entered a county. Only those who owned two hundred acres of land could be elected to the legislature, and the governor must possess a $500 freehold estate. The machinery of government—executive, legislature, judiciary—was similar to that of most other states. But, as part of its "Bill of Rights," the convention decreed that the equal participation of the free navigation of the Mississippi River "is one of the inherent rights of the citizens of this State. . . ." Let Spain, crouched at the mouth of the Mississippi, take heed!

Jackson's part in these proceedings was minor. Tradition, in lieu of something better, credits him with suggesting the name of the state: Tennessee, derived from Tinnase, the name of a Cherokee chief. But so much of the "tradition" surrounding Andrew Jackson has otherwise proved apocryphal that the skeptical mind automatically rejects such an excellent story. What Jackson did do at the convention was to second the motion by which the legislature was designated bicameral, and he was one of the two members from Davidson County to sit on the committee to prepare the initial draft of the constitution. When the draft was first submitted, a member on the floor proposed that a profession of faith be required of all officeholders—that they believe in God, heaven and hell and the divine authority of the Bible. Jackson, along with most of the prominent men of the convention, objected to the motion and they succeeded in striking out the last clause. But if Jackson's action on this point appears liberal, he switched to the more conservative side on another issue dealing with religion. This one involved banning clergymen from the legislature and from any other civil or military office or place of trust within the state. Jackson, disapproving such a far-reaching proscription, seconded a motion to limit the ban to the legislature, and this eventually carried.

Whereas Jackson's role in the convention was thus modest, it was nonetheless respectable, and it suggested that he had an intuitive sense for the middle ground. His voting oscillated between a slightly liberal to a more pronounced conservative position. Acting instinctively, he associated himself with the largest number of delegates within the convention. Although he was not given to any particular philosophy of government that approximated the thinking of others, he knew how to cast his votes wide enough to encompass many diverse segments of the group. All of which indicated he was developing rapidly as an able politician.

With the constitution written and approved, Tennessee was admitted as the sixteenth state in the Union on June 1, 1796. A legislature was elected, John Sevier was chosen the first governor, and William Blount and William Cocke were elected United States Senators. To counteract Sevier's popularity, the Blount faction needed an attractive candidate to run for Tennessee's single seat in the House of Representatives, but they also wanted someone they could depend upon, someone who understood the value of party cooperation. They chose Jackson. For his part, the young attorney-general accepted the offer because he was anxious for advancement and because he knew the Blount clique had leadership, organization, and money. The Sevier group lacked these essentials and relied mostly on the popularity of its leader. Besides, Jackson was already indebted to the Blounts for his present office as attorney-general and even if he contemplated switching his allegiance (which was unlikely), he could expect little consideration from the Sevier group. So he threw in with the Blount crowd, agreed to run, and was elected in the fall of 1796 as Tennessee's only representative in the lower house of Congress.[5]

When Jackson left for Philadelphia, then the nation's capital, it was not his first trip to the great city. He had been there in 1795 to sell land: fifty thousand acres that he held jointly with John Overton, and eighteen thousand acres on commission for Joel Rice. After twenty-two days of what he called "difficulties such as I never experienced before," Jackson at last found a buyer in David Allison, a Philadelphia merchant and speculator. Allison bought the lands at a fifth of a dollar per acre and gave notes to cover the entire amount of the property. Jackson, who planned to open a trading post on the Cumberland River in Tennessee, took his share of the money and purchased goods from Meeker, Cochran and Company, endorsing many of Allison's notes to the company to pay for the supplies. Then, in the fall of 1797, Allison went bankrupt. Meeker, Cochran and Company notified Jackson that Allison had defaulted and that he, Jackson, would have to make good for the notes he had signed. To cover himself, Jackson sold his store for thirty-three thousand acres of land; then he sold the land for a quarter of a dollar per acre for which he received a draft on William Blount, his friend and political mentor and supposedly a rich man. Hurriedly, he returned to Philadelphia to cash the draft, only to discover that Blount himself was involved with Allison and was now caught in a financial squeeze because of Allison's failure. Jackson was nearly beside himself when he learned of this latest debacle, but there was nothing he could do except pay what he owed or face ruin and prison. Thus, year after year, as the notes came due, Jackson spiraled deeper into debt as he essayed one speculation or financial deal after another in an effort to disentangle himself from his predicament. He finally sold his plantation at Hunter's Hill and took up residence in a log cabin at the Hermitage. In so doing, he learned a lesson he never forgot. Thereafter, he regarded paper money and debts as the instruments of the swindler and the cheat, instruments to defraud the innocent and the honest. For Andrew Jackson, hard money—specie—was the only legitimate money, the kind a man could put between his teeth and bite.

His return to Philadelphia as a congressman in 1796 would hopefully end in better circumstances. He arrived just as the nation concluded its first contested presidential election. George Washington was about to complete his second term in office, would retire to his Virginia home the following spring, succeeded by John Adams, the candidate of the Federalist party, who won the election by a majority of three votes over his opponent, Thomas Jefferson, the candidate of the Republican party, who was chosen Vice-President instead. As Jackson submitted his credentials of election to the Speaker of the House there was a look about him that was unmistakably Western. Albert Gallatin of Pennsylvania, Swiss born and French accented, remembered the sight and later described Jackson as "a tall, lank, uncouth-looking personage, with long locks of hair hanging over his face, and a queue down his back tied in an eel skin; his dress singular, his manners and deportment those of a rough backwoodsman." Gallatin could never abide this Western upstart and his overwrought description has more malice than truth in it. To depict Jackson as uncouth, ill-mannered and dressed in a bizarre costume is a gross libel. There surely were rough spots in Jackson in 1796, but he had been around and he knew how gentlemen dressed. As one of the "first" men in Tennessee he would no more walk into Congress outlandishly rigged than he would allow some poltroon to insult his wife.[6]

On the third day of the session Jackson heard Washington

deliver his farewell address to the Congress. When the great man finished, a committee of five was appointed to write a reply, a customary formality at that time. The committee returned with a glowing tribute to the retiring President, and a vote was called for its adoption. Jackson voted nay. He emphatically rejected any salute or gesture of respect for the President because he believed that the Jay Treaty with Great Britain, concluded by the Washington administration, was a stain on the honor of the Republic. Also, he believed President Washington guilty of enforcing Indian treaties that abrogated the claims of white settlers. Later, when he ran for the presidency himself, this vote caused Jackson great injury, since it was argued that anyone showing the slightest disrespect to the Father of his Country did not deserve to sit in the White House.

Jackson's only other action of importance during his brief career as a congressman was his introduction of a bill to reimburse Tennessee for the expedition of John Sevier against the Cherokee Indians in 1793. He had to fight a legislative battle to win his case, but after a speech in which he successfully conveyed some of the horror of tomahawk warfare, especially as waged against women and children, the House approved his proposal and the Senate followed along shortly thereafter. Tennessee received over $20,000 as reimbursement. Jackson's other actions in the House, particularly his voting record, revealed strong nationalistic tendencies; for example, he approved the completion of three frigates and disapproved the practice of bribing the Barbary pirates to permit U.S. ships to sail through the Mediterranean Sea unmolested.

When the people back home read the record of Jackson's performance as a freshman congressman—especially in obtaining the $20,000—they felt a warm glow of pride in his accomplishments. He won their approval so quickly that when William Blount was expelled from the Senate for scheming to help the British in an attack upon Spanish

Florida in order to prevent the possible closing of the Mississippi River, Jackson was elected to replace him. But this was a mistake. Jackson was out of his depth as a Senator. And, after taking his seat in the upper house, he paid little attention to his duties. He was distracted. His mind was constantly engaged by his financial problems caused by Allison's default. His letters bemoaned his predicament. "If he [Allison] was only possessed of honesty," Jackson wrote ruefully, "but this is wanting. I happened to be security for his appearance at Jonesborough in '88. Judmt, last court was passed against me as his Bail for upward of Two Hundred Dollars, and D—n the Rascal, he will not Evan convey me land to the amount." To escape punishment, Allison finally mortgaged eighty-five thousand acres of Tennessee land to one Norton Pryor; but it was too late. He was sentenced to debtor's prison, where he died not long after. To make matters worse, the panic of 1797 struck the country a glancing financial blow, but it had enough wallop in it to send Jackson reeling.

With his financial problems escalating, Jackson asked for and obtained a leave of absence from the Senate in the spring of 1798. Then, almost as soon as he reached home, he resigned his office. A shabby business! While he was unsuited for the Senate, and distracted by the Allison disaster, it was not for these reasons that he resigned. What probably inspired his resignation was his desire for appointment for another office, one that would keep him close to his interests in Tennessee, one that would pay more money than any other office in the state (with the exception of the governorship), and one that would advance his political ambitions by taking him to all parts of Tennessee, bringing him into immediate contact with the people. What his hard eye gazed at now was an appointment to the Superior Court, a post that carried a salary of $600 a year. With his impeccable political record, he had no trouble with the legislature, and he won election almost on signal. He served in this post for six years, most often sitting in Nashville, Knoxville, and Jonesborough. In the course of riding the circuit, he met most of the leading lawyers and politicians in the state. Because he was a man of innate dignity and bearing, he wore a long black grown in order to look like a judge when presiding. However, his opinions did not always match his judicial mien and one man described them as "short, untechnical, unlearned, sometimes ungrammatical, and generally right." If they were indeed generally right, then surely justice was never better served, with or without the niceties of legal jargon.

His retreat to the bench prompted Jackson to renew his effort to settle the Allison business. He prevailed upon Norton Pryor, who had received the mortgage for eighty-five thousand acres, to join him in a suit against the Allison heirs so that they could gain a clear title to the land. Because Jackson was a Superior Judge and could not be involved in the proceedings, he employed his friend, John Overton, to take his case. Overton instituted suit in a Federal court and won. Jackson received five thousand acres as his share of the settlement and promptly sold the land. Then, years later, he learned that the decision was invalid because a Federal court lacked jurisdiction in such cases. Jackson was therefore liable to suit from those to whom he had sold the five thousand acres—and not merely for the original value of the land, but for its current value! Total disaster faced him. Without wasting a moment's time, he galloped off to Georgia where he understood the Allison heirs now lived. He found one member of the family, a William Allison, who, said Jackson, "knew the Justice of my claim, and how much I had suffered in sacrafice of property" because of the original bankruptcy. He prevailed upon Allison to convene the other heirs, and with a persuasion born of desperation he convinced them to sign over to him all their property rights in Tennessee. The release safely tucked away in his pocket, Jackson returned home and granted a clear title to those who had originally bought the five thousand acres from him. It was a close call.[7]

During his term as Superior Judge, Jackson not only had trouble with the Allison affair, but he also fell into a fearful row with Governor John Sevier. Sevier was an easygoing, affable, generous man, a popular Revolutionary War hero and Indian fighter who, when he appealed to the people for their votes, was practically invincible. But he did not have a political organization, nor did he understand its value and importance. He left everything to his popularity—and sometimes it was not enough.

When Sevier's current term as governor ended, he was ineligible for re-election and was succeeded by Archibald Roane, one of the young lawyers who had come West with Jackson. Shortly afterward, in 1802, the office of major-general of the militia became vacant, and both Sevier and Jackson decided to go after it. Although lacking in military training or experience, Jackson had been maneuvering to acquire this office for some time. For one thing, he appreciated its value in promoting his reputation and advancing his career. In 1792, he made a stab for it, and though the Blount faction was willing to gratify his wish, the arrangement did not work out. He tried again in 1796, and again he failed. In 1802, he was determined to make an all-out effort and carefully prepared his strategy. Under Tennessee law, the office was filled in an election by the field officers of the command. Employing the skills of political organization he had learned from the Blounts, as well as relying on his own talents, he conducted a brilliant campaign. Despite the fact that he was running against a popular war hero, he succeeded in bringing off a tie. The rest was easy. Governor Roane stepped in and broke the tie by voting for his friend. Thus, at the age of thirty-five, Andrew Jackson became a major-general of the militia.

Sevier was livid with rage. That a lawyer, speculator, and politician would dare to vie for military office with a man whose military experience included the great victory of King's Mountain during the Revolution seemed incredible to Sevier—nay monstrous—and his friends, in revenge, jammed through the legislature a bill stripping Jackson of part of his command by creating two militia districts, one in the east and one in the west, and permitting Jackson to retain only the western command. A year later, Sevier was again eligible for the governorship, and he ran against Roane who was seeking a second term in office. Jackson, naturally, came to the aid of Roane. What ensued was a bitter campaign of personal vilification and insult, with Jackson furnishing Roane information charging Sevier with land frauds. The situation finally erupted into a brawl when Jackson came to Knoxville one day to open court and happened to pass a crowd being harangued by Sevier. Seeing the upstart, Sevier poured out his wrath at him, sneering at his pretensions. Jackson defended himself by referring to his services. "Services!" laughed Sevier, his voice soaring with sarcasm. "I know of no great service you have rendered the country, except taking a trip to Natchez with another man's wife."

"Great God!" screamed Jackson, "do you mention *her* sacred name?"

Shots rang out in the crowded street, as though to punctuate Jackson's fury. However, before a general riot could begin the two combatants were borne away by their friends. Jackson immediately challenged Sevier, but the older man put him off. "You without provocation made the attack," wrote the General, "and in an ungentlemanly manner took the sacred name of a lady in your poluted lips, and dared me publickly to challenge you. . . . As your age protects you from that chastisement you merit, the justice I owe to myself and the country urged me to unmask you to the world in your true colors; in the Gazette of Monday next. . . . The advertisement as follows, 'To all who shall see these presents Greeting. Know ye that I Andrew Jackson, do pronounce, publish, and declare to the world, that his excellency John Sevier . . . is a base coward and poltroon. He will basely insult, but has not courage to repair, Andrew Jackson."[8]

The incendiary words were published just as Jackson promised. Sevier could no longer refuse the challenge and agreed to meet Jackson across the border in Kentucky. But things went askew. Instead of two men engaging in a serious duel to the death they inadvertantly got entangled in low comedy. It was a wild episode, with Sevier arriving late for the appointment, with the mounted Jackson charging at him with a drawn sword cane like some medieval knight, with frightened horses running away with the pistols, with Sevier ducking behind a tree to escape injury, with the seconds drawing pistols on each other to keep the fight fair, and with lord-only-knows-what-else, certainly nothing approved by the strict etiquette of the Dueling Code. Finally the mad men were prevailed upon to cease their murderous quarrel, and the whole party rode back together to Knoxville. Sevier won his election and went on to complete a second series of three administrations as governor. Fortunately for Jackson, the two events most important to his military career—his election as major-general of the militia, and the outbreak of the War of 1812—came at times when Sevier was out of office and had been replaced by someone from the Blount faction.

This duel with Sevier increased Jackson's reputation as a man of honor, fearless and brave, who would take abuse from no one, not even a former governor and war hero. But perhaps his most famous gunfight was the one that occurred in 1806 with Charles Dickinson, reputedly the best shot in Tennessee. The affair grew out of a horse race wager. Jackson's horse, named Truxton, was matched against

Ploughboy, owned by Captain Joseph Erwin. The stakes were set at $2000 and the forfeit at $800. The day before the race was scheduled to take place, Ploughboy went lame and was withdrawn. Erwin and his son-in-law, Charles Dickinson, paid the forfeit. About this time, Jackson also experienced some grave unpleasantness with Dickinson who, while drinking, had several times dared to mention the "sacred name" in public and in a manner that was far from reverent. Jackson confronted the twenty-seven-year-old dandy. Dickinson apologized, offering as his excuse the fact that he had been drinking and did not know what he was saying. To complicate matters in the interim, the notes offered by Erwin to pay the forfeit were declared different from those agreed upon at the time the wager was made. Dickinson denied this, replying (so it was reported) "with abuse." Hot words were exchanged, and Jackson issued his challenge. But this was not a sudden decision on Jackson's part, born of passion and anger. It was something that had been brewing for a long time, and the challenge came only because there was no other way to shut Dickinson's mouth.

The two men met on a late May day in 1806. Eight paces (24 ft.) were measured off. "Are you ready?" asked the second, John Overton. "I am ready," replied Dickinson. "I am ready," responded Jackson.

"Fere!" called Overton in his old country accent.

Dickinson raised his pistol quickly and fired. The ball struck Jackson in the chest and as it hit him, a puff of dust rose from the breast of the coat. Jackson stood perfectly still, then raised his left arm and pressed it tightly against his throbbing chest. "Great God!" cried Dickinson, "have I missed him?" As Dickinson waited for the return shot, Jackson took his time as he slowly squeezed the trigger. With a click the hammer sprang forward but stopped at half-cock. Jackson drew it back, aimed a second time, and fired. Dickinson took the shot just below the ribs. As he dropped to the ground his assistants rushed to his aid and tried to stop the rush of blood. But they could do nothing. The unfortunate man bled to death.

Jackson had been saved by the loose-fitting coat he wore. Because of it, Dickinson misjudged his target. Instead of piercing Jackson's heart, his bullet fractured a rib and raked the breast bone. Still the wound proved difficult to treat and it was almost a month before the General could move around. The doctors did not dare probe for it and he carried it in his body until his death.[9]

Jackson responded well to the devotion and excellent care of his wife. He recovered from the Dickinson fight at his home on the estate he called the Hermitage. This was not the great mansion he later built; this was a two-story block-house, a modest yet comfortable home with one room on the ground floor and two rooms upstairs. A short time after-ward, he built a small structure about twenty feet away and connected it to the main house with a covered passageway. When he built the famous Hermitage mansion in 1819, he converted the older dwelling into a one-story Negro cabin. And, after 1804, he had a great deal of time to build—even-tually he had six homes—to tend the affairs of the plantation and to look after his other business interests. For one thing, he resigned as judge of the Superior Court in 1804. He had had enough of the bench and was ready for something more challenging. With his personal finances slowly beginning to improve once the Allison business was disposed of, he could afford to indulge himself and so he pitched his ambitions to nothing more or less than the governorship of the Orleans Territory, recently sliced out of the southern portion of the Louisiana Purchase. Ultimately, he failed in his quest, but what is fascinating about this unsuccessful reach upward was the manner in which he campaigned for the job. Over the years, due in part to his association with the Blounts, he had acquired an education in the dynamics of politics. Thus, when he applied for the governorship, he sheathed his ambitions in the requests of others. Among other things, he gathered the entire Tennessee delegation in Congress to his cause and had them sign a petition on his behalf to President Jefferson. This was a very respectable accomplishment and took care and work; but he was under no delusions that his campaign would succeed without offering such a petition decorated with the names of loyal and important Republicans. Next, to provide additional thrust, he arranged a personal appeal to Jefferson from Matthew Lyon, one of the martyrs of the Republican party who had been sent to prison for violating the Sedition Law during the John Adams Administration. The net result was a well-prepared and convincing case. Unfortunately, it failed to convince Jefferson, and the appointment went instead to W. C. C. Claiborne. However, what is significant here is not his failure or his soaring ambition, but those small touches in preparing his application that indicated Jackson's developing skill as a politician. Of course, all his life, he liked to imagine that he did not pursue public office, that it pursued him in recognition of his manifold services to the Republic and the American people. The truth of the matter is that like all successful politicians he was first and foremost a man in imperious pursuit of his star who would never be so foolish as to rest his career in the uncertain hands of the masses.

During the next few years Jackson continued to talk and act like a Jeffersonian. But he was completely soured on the third President—the consequence of his failure to obtain the Louisiana post. As he swung away from Jefferson he veered closer to the more extreme States' rights wing of the Republican party and took his political direction from such men as John Randolph of Virginia and Nathaniel Macon of North Carolina, two doctrinaire States' righters. By 1808, Jackson's hostility toward Jefferson was so pronounced that he was regarded by many as the leader of the movement in

Tennessee to run James Monroe for the presidency in place of the party choice, James Madison. Yet, while it is true that Jackson's politics could be shaped by something as petty as the deference shown him by public men, it is also true that there was a conservative States' rights streak in him that showed itself at the most improbable times.[10]

After his failure to win appointment as governor, Jackson returned to his private affairs. He was somewhat active in the operation of the local Masonic lodge, for together with such men as George W. Campbell, John Rhea, William Dickson, Jenkin Whitesides, and others he organized a Masonic lodge in Greenville, Tennessee, on September 5, 1801, acting under a "Dispensation" of the Grand Lodge of North Carolina. But his business concerns consumed most of his time, and fortunately he could rely on the managerial skill of his wife, Rachel, to help him improve the value and productivity of his land. Theirs was one of twenty-four plantations in the county that boasted a cotton gin. The gin not only serviced Jackson's crop but that of his neighbors as well—for which the General took a fee in money or goods. He also had a distillery that regularly produced several hundred gallons of whiskey, but it burned down in 1801 and all the copperware, stills, caps, worms, and the rest of the equipment were destroyed. To work his plantation Jackson started with a few slaves which he increased in number through the years until he owned approximately 150 Negro workers. On the whole he treated his slaves decently but when he felt punishment was needed, he had them whipped and, on occasion, chained. Under the care and attention of this workforce, the land yielded a variety of crops, the most profitable being cotton, corn, and wheat. Domestic animals were also raised on the plantation such as cows, mules, pigs, and of course horses. One of Jackson's great passions in life was the breeding and training of horses, good horses, racing horses like Truxton.

When he left the bench, Jackson found a great deal of

time on his hands. In the hope of realizing a respectable profit, he returned to storekeeping. He opened a general merchandising establishment and took in two partners to assist him in its operation: John Coffee and John Hutchins. Coffee had some experience in the trade and later added to his advantage by marrying one of Rachel's innumerable nieces; while Hutchins already merited distinction because he was one of Rachel's favorite nephews. The firm was located at Clover Botton, the race course, which was about four miles from the Hermitage and seven miles from Nashville where the Lebanon Road touched the Stone's River. It was nothing more than a blockhouse at first, which Coffee also used as a residence. Later, Jackson added a tavern, stables, and outhouses making it a favorite meeting place for local society. The store obtained most of its things from Philadelphia and sold such items as cloth, blankets, calico, grindstones, coffee, rum, hardware, gunpowder, salt, and cowbells—all, naturally, at triple the Philadelphia price. Rarely did the company get money for its goods, receiving instead, cotton, ginned and unginned, wheat, corn, tobacco, pork, skins, and furs. These items were then floated down the Cumberland, Ohio, and the Mississippi rivers to Natchez, where they were sold in the New Orleans market. As a side line, because of its excellent location on the Cumberland River, the firm also built river boats for other traders.

But before long Jackson proved again he was no man for business. The store failed. There were several reasons for the failure, including a succession of bad debts and a breakdown of communication between Nashville and the lower country, which made it difficult to figure prices and transportation costs. With proper communications gone, the company lost money rapidly. Jackson sold out to Coffee, taking notes payable over long intervals. Soon afterward Coffee quit the business and returned to surveying. However, when he married Rachel's niece, Polly Donelson, he was deeply touched as Jackson, on the wedding day, took the notes out of his strong box, tore them in two, and, bowing to Polly, handed her the fragments.[11]

Jackson was more successful in breeding horses. When he resigned from the bench, he immediately set off to Virginia to find the most perfect horse in the country and bring him back to the Hermitage. Truxton was the result of that trip; and not only did Truxton win races that carried several thousand dollar bets on his nose, but he also earned high stud fees.

Thus, Jackson's sources of income were varied. Apart from his law activities, he earned money from the plantation, the cotton gin, horse breeding and racing, and—it should be noted with regret—from an occasional fling at the highly lucrative business of slave trading. Probably he got involved in this traffic because he had agents in Natchez to whom he sent a steady stream of boats when he operated the store. Since the boats were loaded with produce of every type, it seemed sensible to ship slaves too. But slave trading was a sordid business and carried a social stigma which no gentleman cared to bear—unless of course the profits were high enough and the stigma concealed from public view. Jackson did not regularly ply this squalid traffic and many times he merely transported the slaves to the lower country as a service for a friend or client. Later he was unjustly pilloried by partisan newspapers for running an extensive slave-trading operation; nevertheless, he merits censure for the few times he stooped to this filthy business.

With so much in the way of material goods, with so much love and respect binding Andrew and Rachel together, it is a great pity the Jacksons never had any children of their own on whom to lavish their prosperity and devotion. Not that the Hermitage lacked the presence of children. Several times Jackson served as guardian for children whose fathers had died. First there were the two children of Edward Butler; then came John and Andrew Jackson Donelson, the sons of Samuel Donelson, whom Jackson raised and schooled after the death of their father. Finally in 1810, Rachel and Andrew were permitted to adopt legally one of the twins born to the wife of Severn Donelson. The child was christened Andrew Jackson, Jr. He brought much joy to the family, particularly to Rachel whose life with her husband was soon punctuated by prolonged separations. The child helped fill her many lonely days. On one occasion when Jackson was off fighting Indians, Rachel wrote him, "Often does [our little Andrew] ask me in bed not to cry, papa will come again and I feel my cheeks to know if I am shedding tears. On Thursday last, he said, 'Mama let's go to Nashville and see if he's there'. I told him where you had gone. He said, 'Don't cry, sweet mama'. You can't think how that supported me in my trials."[12]

Then, into this happy and prosperous home one bright May day in 1805, the socially prominent Andrew Jackson brought a distinguished visitor. It was none other than Aaron Burr, the former Vice-President of the United States and slayer of Alexander Hamilton. Burr had unsuccessfully tried to slip past Jefferson and take the presidency in the election of 1800–1801. In Western eyes, that bit of highjinks was more than redeemed by his accurate shooting at Weehawken. Presently, this agreeable and delightful man was sporting himself out West and dropping hints that some delectable tragedy was about to engulf the hated Spanish in the Southwest. People came from miles around to see him, invite him to their homes, and bask in his obvious brilliance. When Burr reached Nashville, he selected the fashionable Jackson family to honor with his presence and stayed at their home for five days. Three months later, he was back with plans to expel the Spanish from the Southwest and colonize the region with Americans. He knew full well that any expedition against the Spanish automatically produced Western sympathy and support, and Jackson, with his fire-breathing nationalism, was an easy mark for Burr's scheme. "I love my Country and government," wrote Jackson at this time. "I hate the Dons. I would delight to see Mexico reduced, but I will die in the last Ditch before I would yield a foot to the Dons or see the Union disunited." It is still not clear—and probably never will be—precisely what Burr was up to as he plotted his Western coup. Possibly he was unsure himself and planned to take advantage of whatever developed. Very probably he could trifle with treason against the United States should that prove personally advantageous; but if treason was indeed a possibility he never divulged it to Jackson.

When Burr returned to the East after his initial swing through the West he corresponded briefly with General Jackson, even to the extent of mentioning the recruitment of troops and their transport to the South—possibly to New Orleans—in the event of war with Spain. Drawing his new ally deeper into his conspiracy, Burr also asked Jackson to provide him with the names of men who could serve as officers for one or two regiments, from colonel down to ensign. With his natural caution in such matters the General consulted his friends in Nashville before responding. After hearing their encouragement and support and after reassessing the almost universal popularity of a Spanish expedition among Westerners, Jackson sent the list of names.

The following summer, 1806, Burr again appeared in the West. Leaving his daughter, Theodosia, at Blennerhassett's Island in the Ohio River, the staging area for the conspiracy, he descended the rivers and arrived at the Hermitage on September 24, 1806. A ball was given in his honor, and although there were whispered rumors that Burr was dabbling in treason, Jackson's hospitality served to stigmatize the rumors as vicious and unfounded. Apparently satisfied that the General was a convert to his schemes, Burr entrusted Jackson with two important orders: one directed him to build five large boats to be used for descending the river, and the other instructed him to purchase provisions. Accompanying the orders was a packet of Kentucky bank notes in the amount of $3,500. During all of this increased activity involving men and boats and money Jackson steadfastly refused to believe that anything improper was happening—because that was what he wanted to believe. He took the money to his partner, John Coffee, and directed him to build the boats.[13]

Not until November 10, 1806 did the cloud of suspicion enter his mind. While Jackson was talking with a friend, one Captain Fort, at the Hermitage, the captain let slip the information that a plot was underway to take New Orleans and divide the Union. Jackson was incredulous. Until that moment, he later swore, he believed that Burr intended only to found a colony near the Red River and, in the event of war with Spain, to march into Mexico. Now, he also learned that General James Wilkinson was involved with Burr, the same Wilkinson who commanded the U.S. troops in Louisiana and whom Jackson knew for a scoundrel. When he discovered this, Jackson's apprehension turned to fear. Still he was a cautious man and unwilling to risk embarrassment by making unsubstantiated accusations. What he did, therefore, was to write a carefully worded letter to President Jefferson, in which he protested his loyalty and offered the services of his friends and neighbors—provided, of course, they were mustered under his command. "In the event of insult or aggression made on our government and country FROM ANY QUARTER," he wrote, "I am well convinced that the public sentiment . . . within the State . . . are of such a nature . . . that I take the liberty of tendering their services, that is under my command. . . ." At the same time Jackson wrote to Governor Claiborne and warned him to "Be on the alert, keep a watchful eye on our Genl. [Wilkinson], and beware of an attack as well from your own Country as from Spain." It was not a direct accusation—just a warning. Yet, to cover himself in the event the rumors and suspicions were false, Jackson also contacted Burr and asked for an explanation of what was going on, assuring him in the same breath that he would never condemn a friend on hearsay evidence alone. Burr responded by claiming that he had the complete approbation of the federal government and that he himself received a blank commission signed by President Jefferson. These were reassuring words and may have convinced the General; more likely Jackson accepted the explanation because he wanted the Spanish ousted from the territory, and especially because he was aware of Tennessee enthusiasm for a filibuster in the Southwest. Furthermore, he habitually judged other men's loyalty according to their personal regard for him. Since Burr had always been deferential, Jackson summarily cleared him of suspicion. Consequently, he delivered the boats as contracted (the order had subsequently been reduced to two) and returned the balance of the bank notes. The General may also have been prodded into carrying out his end of the deal by the nagging fear of legal action if he defaulted.

Burr and sixty followers sailed off in the boats on December 22, 1806. They were not very far down the river when word was received that a Proclamation had been issued by President Jefferson to the effect that a military conspiracy existed in the West. Acting on information provided by the turncoat Wilkinson, the President ordered proper authorities to apprehend the guilty. Burr was picked up as he fled toward Mobile and was brought to Richmond, Virginia where he was tried for treason. Unfortunately, the trial degenerated into a political duel between the presiding judge, John Marshall, and President Jefferson. As a result, Burr walked away acquitted.

Jackson was later to suffer great political damage because of his part in the conspiracy. With good reason, too. Not only had he delivered the boats, but at Richmond, where he was summoned to give testimony during the trial, he told a sympathetic crowd that he believed Burr innocent of the charge brought against him. It may be that the speech was intended primarily for Western ears. Or, it may have been a typical display of Jacksonian loyalty, or a senseless indulgence of his dislike for Jefferson. In any event, both Jefferson and his Secretary of State, James Madison, took a dim view of his opinion, and Madison later remembered it when he sought generals to fight the war against Great Britain.[14]

About twenty years after the trial, during the presidential race of 1828, Jackson was accused of having played both sides for his own advantage, that he pandered to Burr at the same time he shielded himself with paper protests to Jefferson and Claiborne, and that he himself had come perilously close to treason if he had not indeed crossed the mark and betrayed his country. The accusations were ridiculous. The possibility of treason by Andrew Jackson is utterly impossible. If he was overly cautious in responding to the rumors and slow in disassociating himself from the Conspiracy, he was guilty of nothing criminal. After all, good politicians always look both ways before they leap.

CHAPTER III

Old Hickory

AT THE OUTBREAK OF THE WAR OF 1812, THE MILITARY EXPERI-
ence of Major-General Andrew Jackson was virtually nonexistent. He was General by virtue of an election, which was
won by a single vote cast by a political friend. Yet, here was
the man who would command the greatest triumph of
American arms over the British in the history of the nation.
Here was the man who would be remembered as the
supreme military commander of this war: the Hero of the
Battle of New Orleans.

No sooner was war against Great Britain declared by
Congress in the late spring of 1812 than General Jackson
tendered his services to the government through Governor
Willie Blount. Willie was William Blount's brother and the
new leader of the Blount faction in Tennessee. He had
recently succeeded Sevier as governor. Together with his
own services, Jackson offered those of the twenty-five hundred Tennessee volunteers he commanded. However,
because he had so little standing with the national government, owing to his collaboration with Aaron Burr, he
received in return a perfunctory letter from Secretary of
War William Eustis, accepting his services but neglecting to
call him to duty. Caged at home, the frustrated Jackson
vented his fury at the "imbeciles" in Washington. He knew
this was his opportunity for fame and glory, and he yearned
to lead his volunteers against the Spanish in Florida as well
as "carry fire and Sword into the heart of the creek nation."
Not until President Madison had been sufficiently frightened by the possibility of an invasion of the country through

New Orleans did he finally request Governor Blount to send fifteen hundred Tennesseans to reinforce General Wilkinson. Whereupon Blount turned to Jackson. Although the governor appreciated Jackson's feeling about Wilkinson, he asked him to accept the command nonetheless. "At a period like the present," replied Jackson, ". . . it is the duty of every citizen to do something for his country." Still, there was a "sting to my feelings" in the way the matter had been handled in Washington, as though he counted for nothing except as a last resort. Even so, he continued, "viwing the situation of our beloved country at present, should your Excellency believe that my personal service can promote its interest in the least degree, I will sacrifice my own feelings, and lead my brave volunteers to any point your excellence may please to order." Forthwith, Blount instructed him to call out two regiments and hold them in readiness for a march to New Orleans in defense of the "Lower Country." Jackson responded by publishing an order for the volunteers to rendezvous in Nashville on December 10, 1812.[1]

It was a brutally cold winter that year. A foot of snow greeted the volunteers on their arrival in Nashville. Fortunately, the efficient quartermaster, Major William B. Lewis, had provided a thousand cords of wood or the men might have frozen to death. On the night of the rendezvous, the General and his quartermaster walked among the troops from dusk to dawn, making certain the men were protected against the cold. At about six o'clock in the morning, the exhausted Jackson entered the local tavern to rest and have something to drink to warm his body. As he walked through the door, he overheard a civilian lamenting about the terrible organization that allowed the massing of troops without providing adequate shelter for them. It was criminal, said the critic, that men froze in the cold outside while officers lounged inside, enjoying the best accommodations in town. No sooner had the civilian finished than Jackson slammed him to the wall with a fiery blast of words. "You d——d infernal scoundrel," he roared, "sowing disaffection among the troops. Why, the quarter-master and I have been up all night, making the men comfortable. Let me hear no more such talk, or I'm d——d if I don't ram that red hot hand iron down your throat."

It was a measure of Jackson's leadership that he personally supervised the care of his men, never resting or thinking about his own discomfort. Whatever ingredient of character it takes to produce leadership, Jackson had it. And his men knew it; indeed, few of them survived who seriously questioned it.

Several days after they arrived, the cold spell broke, and the troops were organized into an army. Colonel John Coffee was given command of a regiment of six hundred and seventy cavalry, and Colonels William Hall and Thomas Hart Benton were placed in charge of two regiments of infantry comprising a total of fourteen hundred men. William Carroll, a young man from Pennsylvania and a favorite of Jackson's, was commissioned brigade inspector. Finally, on January 7, 1813, the army was put in motion: the infantry sailed down the river in flat-bottom boats, while the cavalry rode south along the Natchez Trace. "I have the pleasure to inform you," the General wrote the Secretary of War, "that I am now at the head of 2,070 volunteers, the choicest of our citizens, who go at the call of their country to execute the will of the government, who have no constitutional scruples; and if the government orders, will rejoice at the opportunity of placing the American eagle on the ramparts of MOBILE, PENSACOLA, and FORT ST. AUGUSTINE, effectually banishing from the southern coasts all British influence."

Trumpeting his nationalism, Jackson announced his readiness for battle. Few others at the time matched his preparedness or eagerness for war. General Wilkinson, for example, distrusted Jackson, and his distrust mounted as the Tennessee soldiers drew closer to New Orleans. He

wanted no part of his glory-hunting, British-hating subordinate and he planned to keep him at a comfortable distance,
away from harm and away from the enemy. He ordered
Jackson to halt at Natchez. Thus, suddenly, the Tennessee
army was jolted to a stop. Weeks passed with no further
orders. Finally, on March 15, 1813 Jackson received a letter
from the Secretary of War ordering him to dismiss his
troops and return home. Jackson was flabbergasted. Five
hundred miles from his base, without pay, without transport, the country at war suffering one humiliating defeat
after another—sent home! Dismissed! At first, Jackson
could not believe it, could not believe that political spite
could carry so far. Worse, here was absolute proof that the
"idiots" in Washington were utterly incompetent to prosecute the war and save the Republic. Best to ignore them,
thought Jackson. Of course he was sure the sly hand of
Wilkinson was buried in this villainous business. He saw
immediately that, because his troops were without food and
shelter, and because they were so far from home, they
would probably join Wilkinson after they were dismissed.
Thus, Wilkinson and the Washington jackals would get rid
of him, take his army, and humiliate him by forcing his
return to Nashville alone. Well, if that is what they thought,
they did not know Andrew Jackson very well. He promptly
sat down and wrote his superior a letter notifying him that
his "brave men at the call of their country, voluntarily rallied
round its insulted standard. They followed me to the field; I
shall carefully march them back to their homes." Instead of
dismissing them he would escort them back to Tennessee. It
was a simple matter of no Jackson, no army.[2]

Although he did not have sufficient money or wagons to
move his army the five hundred miles through the wilderness, Jackson had the pride to get them home somehow,
even if it meant carrying them on his back every foot of the
way. In organizing the return march, he learned that there

were over one hundred and fifty men on the sick list and only a few vehicles to ferry them. Obviously it would be a painful journey of retreat. He ordered his officers to turn their horses over to the sick, and he himself surrendered his three prize horses to the needy and trudged along on foot the entire distance. Jackson's concern for the safety and comfort of his men was genuine enough—"it is . . . my duty to act as a father to the sick," he told his wife—but he had the good sense to conceal it. Still the more he veiled his anxiety the more he pleased his men. During the homeward march a sick boy periodically raised himself from his stretcher to ask, "Where am I?" "On your way *home!*" barked Jackson, as though sounding a call to battle—whereupon the soldiers cheered. As the army caterpillared northward, painfully pulling itself along, there were repeated cheers for the proud commander who silently shared the misery and disappointment of his men. Soon the troops were commenting on how tough their general was, how strong willed and determined. As tough as hickory, they agreed, which was about as tough as anybody knew. Not much later they started calling him "Hickory" as a sign of their respect and regard; then the affectionate "Old" was added to give Jackson a nickname— Old Hickory—that admirably served him thereafter throughout his military and political wars.

It was May, 1813, when Jackson reached home with his bedraggled troops. And no sooner did he arrive in Nashville than he became embroiled in an unseemly affair of honor. The circumstances developed during the homeward march, when a man named Littleton Johnston declared himself insulted by William Carroll and challenged him to a duel. Billy refused on the ground that the challenger was no gentleman. Other men presented themselves as substitutes, but Billy would not give satisfaction. Finally Jesse Benton, the brother of Colonel Thomas Hart Benton, issued a challenge, and his social standing was such that Billy could not refuse without being labeled a coward. By this time, the army was safely back in Nashville so Billy went out to the Hermitage to ask Jackson to act as his second. Quite properly, Jackson refused. Too old, he said. He was forty-six. But Billy claimed there was a conspiracy to run him out of the country, for what other reason could possibly explain the many surrogates? At that, Jackson reared up. "Well Carroll," he said, "you may make your mind easy on *one* point: they sha'n't run you out of the country as long as Andrew Jackson lives in it." To prevent the duel the General spoke to Jesse and succeeded in talking him out of the fight, but later Jesse's friends got him to change his mind again. Disgusted, Jackson consented to act as Carroll's second, and what took place was one of the strangest duels ever recorded in Tennessee history. The details are not clear, but apparently the two men took a back to back position (to give the fight a French twist) and at a given signal strode away from each other, wheeled, and fired. Jesse Benton fired first, then panicked and tried to duck away by turning and bending over. Billy then raised his gun, fired, and hit poor Jesse right in the seat of his pants. Though not lethal, the wound was very painful and very embarrassing. When Colonel Thomas Benton heard that General Jackson had been party to his brother's humiliation he wrote his former commander a reproving letter for conducting the duel in a "savage, unequal, unfair and base manner." Worse, he "inveighed bitterly and loudly, in public places" against Jackson, "using language such as *angry* men did use in the western country fifty years ago."[3]

In public places! Jackson, who was a fierce guardian of his fame and reputation and extremely sensitive to publicized injuries, swore he would horsewhip Thomas Benton for the insult the first time he saw him. It came soon enough. The Benton brothers were in Nashville one day when Jackson and John Coffee came to town and were walking to the post office to collect their mail. Suddenly Coffee spotted

Thomas standing in the doorway of his hotel. "Do you see that fellow," queried Coffee. "Oh yes," replied Jackson, "I have my eye on him." As always in such matters the General was in full control of himself; he did not impulsively rush at Benton to make good his threat; he continued on to the post office to give himself time to think, all the while analyzing whether it was wise to discharge the business now or wait for a less conspicuous place and time. But finally he decided Benton had *publicly* insulted him so he had no choice but to thrash him publicly. The time was now. Jackson collected his mail, turned around and started back to the hotel. Both Bentons were waiting. The General lunged at Thomas, crying, "Now you d——d rascal, I am going to punish you. Defend yourself." Thomas reached into his pocket for what Jackson believed was a pistol. The General drew his own gun and backed Thomas into the hotel. Jesse, meanwhile, scrambled inside, and when the two combatants went by the door he fired two bullets and a large slug straight at Jackson. Old Hickory slumped to the floor, his shoulder shattered by the slug and his left arm pierced by a bullet that buried itself near the bone. Now Coffee rushed into the room, joined by three other men. One carried a sword cane; the others, knives. The swordsman went after Jesse to run him through, but the point of the weapon hit a button and the blade broke in two. The fight ended when Thomas was sent crashing backward down a flight of stairs at the rear of the hotel as he attempted to ward off the knife thrusts of his assailants.

Bleeding profusely, the General was picked up and carried to a room in the Nashville Inn. Several doctors worked over him furiously to stop the flow of blood and save his life. For a time, it looked as though amputation of his arm would be necessary, but the determined Jackson issued one final order before falling into unconsciousness. "I'll keep my arm!" he said.

Jackson kept both arm and bullet. The doctors made no

attempt to remove the missile, and Jackson sheltered the souvenir for nearly twenty years. The wound was dressed with poultices of elm and other wood cuttings as prescribed by the Indians, but it was many long weeks before Jackson could venture from his bed.

For the Bentons, Nashville was no longer a healthy place to live. "I am literally in hell here," wrote Thomas. "My life is in danger . . . for it is a settled plan to turn out puppy after puppy to bully me, and when I have got into a scrape, to have me killed somehow in the scuffle. . . ." So, like the sensible man he was, Benton resigned his commission in the army at the close of the war and headed west to Missouri. Ironically, he was later elected senator from Missouri and became Jackson's ardent political supporter. Then, in 1832, when the bullet was finally extracted from the General's shoulder by a Dr. Harris and proved to be a "half ball of the ordinary size: a good deal flattened by the contusion upon the bone and hackled somewhat on the edge," a member of the family, undoubtedly in jest, offered the souvenir to Benton, supposedly its rightful owner. The Senator declined, remarking that Jackson had acquired title to it in common law by twenty years' possession. It was only nineteen years, however. "Oh, well," replied Benton, "in consideration of the extra care he has taken of it—keeping it constantly about his person, and so on—I'll waive the odd year."[4]

While Old Hickory was recovering from this disgraceful affair, nursed once again by the beloved Rachel, word reached Nashville of the Indian massacre at a place called Fort Mims in Alabama, then a part of the Mississippi Territory. The Creeks had been lashed to war by that remarkable Indian chieftain, Tecumseh, a Shawnee, who plotted to organize the northern and southern tribes into a great confederation and then hurl it against the white man, driving him across the mountains and into the sea. The Creeks, called Red Sticks because they painted their war clubs a bright red color, screamed their approval of Tecumseh's war and ignited the southern frontier with their fury to kill the whites. The settlers fled to Fort Mims. Actually the fort was the residence of one Samuel Mims who had built his stronghold by enclosing an acre of land with upright logs pierced by five hundred portholes and two heavy gates. The Red Sticks attacked the Fort on August 30, 1813, gained entrance, and killed and scalped over four hundred white settlers.

The news of the massacre burst like a thunderclap across the western country. The Tennessee legislature promptly empowered the governor to call volunteers. Blount summoned Jackson to command, and by early October the pale, weak, and unsteady Jackson, his arm bound in a sling, took the field and marched toward Alabama. He pushed his men relentlessly in order to catch the Red Sticks, chastise them and teach them a lesson they would remember always "in bitterness and tears."

As he penetrated Indian country, the General unleashed a widespread assault of savagery against the Indians, killing, burning villages, and plundering food supplies. He met a large force of Red Sticks at the Creek village of Talluschatches, and, after a bloody encounter, systematically slaughtered them. He later boasted that the victory revenged the massacre at Fort Mims. "We shot them like dogs," agreed Davy Crockett. When the battle ended a dead Indian mother was found on the field still clutching her living infant. Jackson asked other Indian women to care for the child, but they refused. "All his relations are dead," they said, "kill him too." The General rejected this solution and afterward took the boy, named Lincoyer, back to the Hermitage and provided him with every advantage, including a good education. Unfortunately, Lincoyer died of tuberculosis before reaching the age of seventeen. This extraordinary action by Jackson— of brutally killing Indian families on the one hand, and extending kindness to an Indian child on the other—was one of several incidents that kept his contemporaries forever perplexed about the range of contrasts fused within this unusual man.

After the Talluschatches victory, many Indian towns wisely declared for Jackson in order to escape his fury. One of these was Talladega. But the Creeks, under the leadership of Chief Red Eagle (also known as William Weatherford, because he was seven-eighths white) attacked Talladega for its treachery and threatened destruction to any other settlement that deserted to the enemy. Aware that future capitulations depended on his ability to protect his Indian allies, Jackson turned his army toward Talladega and sped to its aid. As he approached the town, he formed his men into three lines, stationing the cavalry on the wings. The entire force then advanced with the two ends curved forward. A vanguard was ordered to move ahead, sting the savages with five rounds of shot and then fall back, drawing the infuriated Indians into the curved arms of Jackson's force. The plan worked perfectly, and the Red Sticks charged into the trap. As they entered, the two ends of the line raced together, and the Indians were caught in the circle. The soldiers then fired at them at point-blank range. For a long time the killing was steady and relentless. Finally, some of the Creeks broke out of the circle and escaped, but they left behind nearly three hundred dead comrades, whose bodies littered the ground. Jackson's dead numbered fifteen.[5]

After the Talladega action, the General returned to his base at Fort Strother. When he arrived he learned to his dismay that the supplies he had previously ordered had not been delivered. There followed ten weeks of waiting, waiting while his army grew hungry and mutinous, waiting as hunger weakened discipline and control. Jackson at last ordered the army to return home; but twelve miles out they met the long-awaited supply train and the ravenous soldiers gorged themselves. Jackson then instructed the men to return to Fort Strother, only to have his command greeted with sullen and angry stares. In defiance, one company moved toward Tennessee and was about to be joined by others when the General mounted his horse and ordered his cavalry commander, John Coffee, to block their path. Snatching a musket, he aimed it at the mutineers and warned them he would kill the first man who took another step toward home. Then he enveloped them in such curses and oaths and blood-chilling screams that the men stood paralyzed in their tracks. They studied his face, ridden with dark intentions, and they knew he would shoot them without so much as pausing for a second thought. Abused and broken, the men, at length, turned around and went back to the fort.

Two other mutinies occurred during this campaign that later had political repercussions. One involved an eighteen-year-old soldier by the name of John Woods, who disobeyed an officer and was summarily shot after a quick trial. The other was more famous because it inspired the "Coffin Hand Bill" in the presidential election of 1828. This incident took place in September, 1814, when a group of men, under the impression that their three-month enlistment had expired, left the camp. But the governor had called them to a six-month enlistment, and this fact had been duly published by Jackson. According to the General, they not only deserted and attempted to enlist two hundred others in their mutiny, but they broke into the storehouse and stole supplies. After the ringleaders were captured, they were tried, found guilty, and shot. This incident, along with Jackson's conduct during the entire campaign, was later mounted in a newspaper attack to prove that the General was a tyrant, butcher, coward, and murderer.

When the regular enlistment period finally did end in December, 1813, many of the soldiers left the army despite everything Jackson could say or do to hold them. They streamed away by the hundreds. They had had enough of starvation, Indians, and Andrew Jackson. Then, just as the commander was about to despair, eight hundred new recruits arrived from Tennessee, and before they could discover the hazards and discomforts of Indian warfare in the wilderness, they were marched into Creek territory, where they had no choice but to fight or be killed.

On January 21, 1814, Jackson camped at Emuckfaw Creek. At about six o'clock in the morning of the next day the Red Sticks began an attack but were eventually driven off. Two days later, they attacked again at Enotachopco Creek. This time, the militia panicked and were about to flee when Jackson rode up and implored them to turn around and fight. "In the midst of showers of balls, of which he seemed unmindful, he was seen . . . rallying the alarmed, halting them in their flight, forming his columns, and inspiriting them by his example." A spark of his valor touched his men, and they rallied to his side. The Indians then broke off the engagement and retired.

That was the low point in the campaign. Thereafter, conditions improved. Governor Blount sent new levies, and even the national government conceded that Jackson was doing an important job by dispatching him five thousand fresh troops. Leaving most of these men to garrison a string of posts along the frontier, the General, on March 27, 1814, marched the rest—about two thousand—to the horseshoe bend of the Tallapoosa River where Red Eagle's main force was located. Jackson meant this engagement to end the war.

The Indians had built their village on the neck of land formed by the looping action of the Tallapoosa River, and across this peninsula they had erected a zigzag breastwork of logs, from five to eight feet high, and running four hundred and fifty yards in length. The banks of the river were fringed with canoes, providing quick escape for the Indians if the battle went against them. Within the sealed compound were nine hundred braves and three hundred women and children, almost all of whom thought they were tucked into a safe and impregnable fortress. Arriving at the scene, Jackson's first action was to send Coffee approximately two miles down the Tallapoosa to cut off possible retreat by canoe. Then he planted two cannons about eighty yards from the breastwork of logs and at 10:30 A.M. commenced firing. The cannon balls pounded against the breastwork but could not rip through. Feeling safe inside, the Indians whooped their derision, responding as though they were watching some ridiculous entertainment. Exasperated, Jackson ordered an assault, and the infantry surged forward. The soldiers fought fiercely at the rampart and finally scaled it. Once inside they picked off the Indians with deadly accuracy, pouring round after round at braves, women and children. The Creeks, surprised and confused by the sudden reversal of the battle, raced back and forth within the compound like mindless beasts, while the soldiers skilfully dropped them in their flight. The carnage continued until darkness erased the targets, for Jackson would not halt the shooting because the Indians refused to surrender. With the ruthlessness essential to a great commander, he ordered their extermination.[6]

The next day, 557 lifeless Indians were counted on the ground, but many others died in the woods attempting to escape. Another two hundred corpses were spotted floating in the river. What prisoners were taken were mostly women and children. Jackson's casualties included 55 killed and 146 wounded.

The ghoulish business of checking through the dead failed to produce the body of William Weatherford, for the great chief was absent from the village when Jackson attacked. Not long afterward, in what was surely a fantastic display of courage and audacity, he rode into the American camp. "How dare you ride up to my tent after having murdered the women and children at Fort Mims," demanded

Old Hickory in his best fire-and-brimstone voice. "General Jackson," replied Weatherford, "I am not afraid of you. I fear no man, for I am a Creek warrior. . . . You may kill me if you desire. . . . I am now done fighting." The soldiers, most of them frontiersmen schooled in the belief that all Indians merited instant killing, pressed for his execution, but Jackson drove them back with a command for silence. "Any man who would kill as brave a man as this," he stormed, "would rob the dead!"

Virtually mesmerized by Weatherford's bravery, Jackson offered the chief some brandy and, after a long discussion, allowed himself to be talked into extending mercy to the Creek women and children by agreeing to feed them. Weatherford, for his part, promised to convince the Indians to leave the warpath and return to peace. That said, the chief departed, eventually to retire to his plantation on the Little River; but his life was in constant danger from the relatives and friends of those who perished at Fort Mims.

So the massacre was avenged, and to Andrew Jackson went the glory of having smashed the power of the great Creek Nation. For his success, a grateful government awarded him the rank of Major-General in the United States Army with the command of the seventh military district, an area which embraced Tennessee, Louisiana, and the Mississippi Territory. Now, fully titled and empowered, Jackson summoned the Creek chiefs to a council to hear his terms of peace. First, he demanded twenty-three million acres of their land, constituting what is now three-fifths of Alabama and one-fifth of Georgia. He wanted not simply indemnity and the termination of the Creek war potential, but the removal of the Indians from the borders of Tennessee, Georgia, and Florida in order to safeguard the frontier. He also required a broad and open road from western Tennessee to the Gulf of Mexico. A line was drawn, and the territory north and east of it was guaranteed to the

Creeks by the United States, except that the United States was granted the right to establish military posts, trading houses, and open roads within the territory.

From the white man's point of view, this was a magnificent treaty, fashioned with wisdom and stern justice. To the Indian, it was a cruel and merciless treaty that articulated their humiliation and eventual doom. It took the Creeks more than a month to suppress their rage and finally sign the hideous document. The Treaty of Fort Jackson, executed on August 10, 1814, thus opened up vast Indian lands to settlers; it also transformed Andrew Jackson into a triumphant hero throughout the South and West.[7]

Old Hickory's success compensated somewhat for the series of disasters sustained in other theaters of the war. The repeated attempts to invade Canada ended in humiliating failure, and with Napoleon Bonaparte now in allied custody, Great Britain was free to concentrate her forces and bring the Americans to heel. Catastrophe was already in the making. The British, under the command of Sir George Cockburn, invaded the Chesapeake, marched on Washington, burned the President's mansion along with the Capitol, shelled Baltimore, and raided Alexandria. Meanwhile, the disaffected in New England prepared for a convention to be held in Hartford, Connecticut, at which place some radicals planned to erect an engine to run New England out of the Union.

To add to the mounting havoc, the government heard rumors of an impending British invasion of the South through one of the Gulf ports. The alarms were quickly relayed to Jackson, who hurried his army to Mobile in time to strengthen and garrison Fort Bowyer, which guarded the entrance to Mobile Bay, just as a British sea and land force prepared to fight its way into the harbor. With Mobile temporarily safe, Jackson then wheeled toward Pensacola, a town in Spanish-held Florida only a short distance away.

Supposedly Spain was neutral. Actually, she permitted both a British fleet to anchor in Pensacola Harbor and British troops to muster on shore. But in marching to Pensacola Jackson was possibly creating a score of future problems. To begin with, he had no authority to invade foreign soil. Nor did he trouble himself to request it since the national government was in flight following the invasion of Washington. Besides, his experience with the bureaucrats of Washington taught him not to look for more trouble than he already had. Then, too, his hatred for Spain was so strong as to give him all the authority he felt he needed. "How long," he wrote the Secretary of War, "will the government of the united States tamely submit to disgrace and open insult from Spain." So he stormed into Florida and attacked Pensacola. The British, rather than jeopardize the over-all plans for the projected invasion of the United States, retired to their ships and sailed into the Gulf, while the Spanish meekly capitulated. Then, fearing the British were headed for Mobile, Jackson abandoned Pensacola and retraced his steps to the city he had left a week before. Once back, he waited ten days for the British to appear, and when they failed to show up he concluded that New Orleans, not Mobile, was the focal point of the invasion. Promptly he turned over command of Mobile to General Winchester of the regular army and raced westward. This rapid movement of a sizeable army, darting back and forth through swamps and wilderness to protect the long coastline of the continent was a brilliant feat of generalship. It took him a little more than ten days to get to New Orleans but finally on December 2, 1814, he reached his destination.[8]

The city of New Orleans is slightly over one hundred miles from the mouth of the Mississippi River. It sits alongside the river in a general area of swamps and bayous and great trees festooned with Spanish moss. To the north and east of the city are two large lakes: Pontchartrain and Borgne. The city rests on a strip of land running between Lake Pontchartrain and the Mississippi River, and is approximately four miles from the shores of the lake.

As soon as Jackson arrived in New Orleans, he took a number of defensive actions. First, he called a conference with engineers to learn the best way to seal the city against invasion. Next, he sent squads of men to fell the huge trees in the bayous and so clog the streams and creeks and other small water routes that finger their way toward New Orleans and which might serve as corridors for the British invasion. This excellent idea was totally negated, however, by Jackson's failure to inspect the completed task, or at least to make certain that the obvious entrances were closed by designating some responsible official to check every important bayou. Then he ordered additional batteries of cannon erected at Fort St. Philip, located on the Mississippi River, about half the distance from New Orleans to the mouth of the river. A fleet of five small gunboats was stationed on Lake Borgne to protect the city from that quarter, although it was Jackson's almost unshakable opinion that if the British showed themselves on Lake Borgne it would be intended as a feint to throw him off his guard, and that the main attack would come at Chef Menteur on Lake Pontchartrain, approximately fifteen miles east of New Orleans.

Within the city itself the defensive force on paper totaled only seven hundred men, of which number about two hundred were absent from duty. To complicate matters, there were justified fears about the loyalty of the French and Spanish inhabitants. Another problem was the pirates who operated out of Barataria Bay and engaged in a lucrative if unlawful trade and had the protection of a group of New Orleans businessmen, one of whom was Edward Livingston, a former mayor of New York City. The chief of the Baratarian cutthroats was Jean Lafitte, a French-born blacksmith, a recognized leader and a man of great courage, who resided in New Orleans. As shrewd businessmen the pirates calculated advantages and disadvantages before offering their services. They turned to the British. But the British laid down two conditions for an alliance: the pirates must cease their attacks on Spanish shipping and they must return English booty already seized. That ended that. So the pirates went to Jackson, who at first refused to consort with these "hellish banditti," as he called them, but later yielded to the petitions of Livingston and a committee of the city's leading citizens, who argued that the pirates were really a decent lot and could render essential service to the city's defenses since they were trained gunners and marksmen. With such recommendations from so many of New Orleans' "first men," Jackson relented and accepted Lafitte and his "hellish" gang.

He also accepted the support of free Negroes. They numbered about six hundred in the city, and many of them were quite well to do. "Our country," wrote Jackson to Governor Claiborne, "has been invaded and threatened with destruction. She wants soldiers to fight her battles. The free men of color in your city are inured to the Southern climate and would make excellent soldiers." But this argument appalled the citizens of New Orleans who complained that if guns were placed in the hands of "men of color" a bloody revolt would ensue. Not that that was their only complaint against Jackson. They resented his restrictions against shipping food to Pensacola, where prices were high and profits enormous, and they especially resented his highhandedness, his readiness to crack heads together to resolve problems, and his intolerance of their customs and slow-moving work habits. Of course they anticipated such problems because they thought Jackson to be a wild man from the forest—rough, gross, and alien to polite society. But he surprised them. At one of the first dinner parties held in his honor at the home of Edward Livingston he walked into the reception room looking every inch the man of fashion and elegance. He also

looked like a commander. His sharp-boned face with its
sharp shaft of jaw bespoke authority. Tall and very erect, his
hair now turning gray and standing nearly as straight as the
general himself, he radiated confidence and self-assurance.
He wore a uniform of blue cloth and yellow buckskin, and as
he entered the room he gestured gracefully to the knots of
people standing on his left and right. He made his way to
Mrs. Livingston, bowed his greeting, and then conducted her
to a sofa and sat conversing for several minutes with her
about how kind and generous she was to give the party in his
honor. The Creole ladies were astonished. During dinner he
talked freely and easily with them, mostly about the British
invasion, but he had the good sense to assure everyone that
they had nothing to fear now that Andrew Jackson commanded the city. When he left, the Creole ladies flashed their
celebrated eyes at their hostess and chorused, "Is *this* your
back woodsman? Why, madam, he is a prince!" And, like any
self-respecting prince, Jackson could be ruthless with his
subjects, especially subjects as temperamental as the Creoles
of New Orleans—a group that frequently defied legitimate
authority just for the sport of it. Thus, on December 16,
because the city panicked on hearing reports of the approach
of the British, Jackson instituted martial law to keep the
inhabitants in control. For this act, he was denounced as a
tryrant and dictator, but he disregarded the abuse and
arrested those who defied his orders.[9]

The first report Jackson had of the British approach came
on the afternoon of December 13. He was told the enemy had
entered Lake Borgne from the Gulf of Mexico. American
gunboats on the lake had already been ordered to avoid a
contest, but as United States boats drew back, the wind died
and the British, with forty-three barges equipped with a cannon each, opened fire on the Americans forcing their surrender. Now Jackson had the tricky job of determining which
approach from Lake Borgne the British would take to attack

New Orleans. He did not know it—and the fault was entirely his own—but when the task of felling trees to obstruct water routes leading to the city was in progress someone neglected to close the mouth of Bayou Bienvenue, and this offered safe transit from the lake to the high ground adjacent to the Mississippi River just below New Orleans. The British eventually discovered the unclogged water passage and barged into it. An advanced party of sixteen hundred men under the command of General John Keane was landed, and had Keane acted quickly and marched straight to New Orleans he might have captured it, for the defenses of the city could not withstand a sudden, concentrated thrust. Fortunately, prisoners told Keane that Jackson had eighteen thousand men guarding New Orleans, and that gave the British officer pause. After waiting six hours, Keane decided such a figure was impossible and resumed his march. Within an hour he reached the plantation of one General Jacques Villeré whose son, Gabriel, was a major of the militia. Because the plantation bordered the eastern bank of the Mississippi River, soldiers had been stationed nearby for defense, but they proved useless. In a rapid maneuver, Kean's men surrounded the plantation and captured its inhabitants. During the action, however, Gabriel Villeré managed to escape by jumping out the window of his house and racing to the river, where he found a boat that took him safely across to the opposite bank and eventually to New Orleans.

The news of the safe landing of the British and their movement to the shores of the Mississippi at a distance of seven miles from New Orleans stunned Jackson. He thought he had taken every precaution to prevent it. Since the felling of trees in the Bienvenue area was the responsibility of Villeré, he was later courtmartialed for negligence. But ultimate responsibility rested with Jackson, not Villeré, so justice was probably served when the major was eventually acquitted. Anyway, Jackson quickly recovered from the shock of the British advance. He summoned his staff and told them, "Gentlemen, the British are below, we must fight them to-night." Hurriedly he boarded the schooner *Carolina* and dropped down to the bank opposite the British encampment, while two thousand troops were rushed to positions close to the enemy. At seven o'clock, aided by the campfires that were lit on account of the cold and which beautifully silhouetted the British targets, the *Carolina* commenced firing at the surprised invaders. A half an hour later the main body of the American militia, commanded personally by Jackson, attacked the enemy and fought them hard for two hours, withdrawing only because darkness impeded troop movements. This initial battle took place on December 23 and ended with 24 Americans dead, 115 wounded, and 74 missing; and 46 British dead, 167 wounded, and 64 missing.[10]

Up to this point, General Jackson had committed a number of avoidable errors and had allowed the British to come perilously close to the city. He was still concerned that Keane's expedition was a mere ruse and that the main force would yet appear at Chef Menteur on Lake Pontchartrain. Not until he was informed that the entire British army had been spotted moving to Bayou Bienvenue did he abandon his mistaken notion and recognize that the invasion was before him. Then he swiftly fortified his position north of Keane's encampment. He formed his defense immediately behind a dry millrace, a four-foot-deep and ten-foot-wide ditch, called Rodriguez' Canal, which ran from the eastern bank of the Mississippi River to a cypress swamp about three quarters of a mile inland. Earthen ramparts were thrown up at the northern edge of the canal, behind which artillery pieces were installed at regular intervals. Meanwhile, the *Carolina,* joined by the ship *Louisiana,* spat cannon balls at the enemy to keep them under cover.

On December 25, Lieutenant General Sir Edward Pakenham arrived to assume command of the British force.

He was a man of ability who had been soldiering since the age of 16 and who was the brother-in-law of the Duke of Wellington, the conqueror of Napoleon. He was given command of this army when its previous general, Sir Robert Ross, was killed at Baltimore. Now thirty-seven, Pakenham had fought gallantly against Napoleon in Europe and fully expected to add to his laurels by crushing the Americans at New Orleans.

When he took command, the British army was fully drawn up below the city. As he surveyed the situation, Pakenham immediately recognized the danger of the *Carolina* on his flank. So on the night of December 26, he had several fieldpieces dragged to the river's edge, along with a furnace for heating shot, and on the following morning, opened fire on the *Carolina*. Two well-placed rounds knocked the ship out of control, yet for another hour the batteries continued to tattoo the hull and superstructure of the ship; then, suddenly, the *Carolina* blew up with a tremendous roar, raining burning fragments of wood and metal in every direction. The ground shook for miles, and it sent a wave of terror through New Orleans. To save the *Louisiana* from a similar fate, the Americans towed the ship out of the range of Pakenham's guns.

For the next three days, the British set about erecting gun emplacements in front of Jackson's lines. Beginning at a point about 700 yards south of the American position and working inland from the water's edge, Pakenham constructed four batteries (17 guns) which were capable of hurling a broadside of 350 pounds of metal. The first battery faced across the river to a gun position set up by Jackson on the western bank and commanded by General David Morgan and Commodore Daniel T. Patterson, the American officer in charge of the New Orleans naval station. Pakenham's other batteries were strung out along the line, behind which two columns of troops were formed, with fusiliers on the right and grenadiers on the left. Meanwhile, Jackson strengthened his own defenses, bringing in hundreds of fresh troops from Kentucky, Georgia, Louisiana, and the Mississippi Territory, thickening the ramparts and placing more cannons in position.

As the two sides prepared for the greatest engagement of the war, across the ocean in Ghent, Belgium, American and British peace commissioners signed the instruments that ended the War of 1812. The date was December 24, 1814. None of the issues that had precipitated the war were resolved, but at least the United States had demonstrated once again its determination to preserve its freedom, no matter the cost or risk.

Thus, while some parts of the world heralded the new year with cries of peace, below New Orleans two lines of cannon faced each other and began the year 1815 with a sustained bombardment. For an hour and a half, the firing was so loud and rapid that the delta shook such "as had never before been heard in the western world." The British fired with trained accuracy. They hit a 32 pounder, damaged the carriage of a 24 pounder, then blew up a 12 pounder, along with two powder carriages. The entire scene was soon enveloped in smoke, and the guns glowed red with the heat generated by the rapid fire. It got so bad that a cease fire was issued in order to cool the guns and let the smoke roll away so the men could see their targets. Very slowly the smoke dissipated. As it rose the Americans looked over the rampart. What they saw at first amazed them and then brought loud cheers to their lips. The British had certainly been skillful in their shooting, but nothing like the Americans. Jackson's men, especially the French and pirates, had battered the British position into a "formless mass of soil and broken guns." Once formidable batteries were now totally destroyed and useless. "Too much praise," wrote Jackson, "cannot be bestowed on those who managed

my artillery." But, in a sense, some of the credit for this
impressive performance belonged to the British, for, to protect themselves, they had circled their gun emplacements
with hogsheads of sugar which proved nearly useless.
Bullets pierced them with ease, and the sharpshooting
Americans had no difficulty in killing the men clustered
around the batteries. For their part, the Americans had
girded their position with bales of cotton, and these
afforded maximum protection. Thus, only 11 Americans
were killed and 23 wounded, compared with 30 British
killed and 40 wounded. The veterans of the Napoleonic War
had been taught a lesson in cannonading by the backwoods-
men of America and their allies, the Baratarian pirates.

Pakenham then hit upon a plan to catch Jackson in a
cross fire. He proposed to move some of his troops across
the river, have them seize Patterson's guns, and then turn
the guns against the Americans at the exact moment the
British main force stormed Jackson's lines. It was a fine
plan, and one he probably should have adopted from the
day he took command. However, precise timing and coordi-
nation were essential if the strategy was to succeed. Also, it
involved difficult problems of logistics, since barges were
needed to transport the men across the river—and the
barges were all tied up at Bayou Bienvenue. To haul these
vessels to the Mississippi meant widening a small canal,
named the Villere Canal, that ran to the bayou. So the
British started digging, while Jackson bravely ignored the
grave danger now facing his army. Instead, he prepared
three lines of defense—all on the same side of the river! The
first was the rampart at Rodriguez Canal, located five miles
from New Orleans; the second was erected two miles closer
to the city; and the third still closer by a mile and a quarter.
Apparently, the General expected to fall back from one position to the next if the battle turned against him.

At the Rodriguez Canal, where the major action would

take place, the rampart averaged five feet high, but its thickness varied. In some places it was twenty feet thick, and in others so thin that a cannon ball could whistle through it without difficulty. Along this wall, batteries manned by Frenchmen, the Baratarians and others were set up in three groups: one facing the river, one at the center, and the last guarding the approach along the swamp. Behind the line, 4,000 men, backed by the cavalry of 230 men, arranged themselves into four groups, and another 250 men were posted along the edge of the woods to guard against surprise attack from the flank. On the extreme left of the line stood the Tennesseans, commanded by Carroll and Coffee. Because Coffee's rifle corps were dressed in long coats that made them look like "full trim quakers," the British were astonished by the sight. They had heard that American Quakers were strange, but these facing them, rifles at the ready, were the "d——st" Quakers imaginable. At the center of the American line were the Kentucky troops, the New Orleans militia, and the Negro regiment; and at the right detachments of the 7th and 44th Regiments. Jackson set up headquarters at McCarty's plantation house, which was behind the center of the line and from whose third story window the entire field could be observed with a telescope. However, when the fighting began, Jackson preferred a closer view near the line, so he moved his staff forward.[11]

On the western side of the river, however, where the American position was extremely vulnerable, little was done to strengthen the command. The force, commanded by David Morgan, amounted to eight hundred men, most of whom were badly armed. Although Jackson guessed correctly that the major British thrust toward the city would come on the eastern bank of the river, he obviously did not understand how easily and decisively he could be beaten, provided the British could capture Morgan's batteries.

On January 7, Commodore Patterson walked along the

western side of the Mississippi to a point directly opposite the British position and there watched the enemy's movements for several hours. What he saw revealed the British plans. Pakenham had deepened and widened the canal and had brought fifty barges to the river, which were to be used to ferry troops to the western shore at nightfall. Jackson was notified and additional aid was rushed to Morgan's command, but it was too little and too late. The Battle of New Orleans was about to begin, and the Americans were dangerously exposed to a possible cross fire.

At six o'clock the following morning, January 8, 1815, the light of the new day dimmed the campfires and was in turn dimmed by a Congreve rocket that rose with a screech and burst over the British position. The rocket signaled the beginning of the Battle of New Orleans, a battle that decisively changed the course of American history. The main attack was planned to begin along the cypress swamps, thought to be the weakest part of the American defense. Prior to that, however, another column was expected to move along the edge of the river to divert Jackson's attention. And a third force, after crossing the Mississippi, was instructed to move up fast and capture Patterson's position. Those in the attacking columns in front of the Rodriguez Canal were ordered to carry bundles of sugar cane with them to throw into the canal to fill it up, then scale the rampart using scaling ladders.

But delays and confusion blunted the assault. The force crossing the river lost three hours in transit; the column nearest the swamp took a cannonading that ripped great holes in the lines and tossed men around like gunny sacks. Finally when the attacking column came within two hundred yards of the rampart the Americans opened up with small arms. Tennessee rifles, Kentucky muskets, four lines of sharpshooters sent a murderous fire into the faces of the British soldiers, who pitched to the ground, falling on top of one another.

Again and again the Americans raked the enemy with shot so continuous that a British officer said the rampart "looked like a row of fiery furnaces!" At a hundred yards from the American position, the British front column hesitated, halted, and then recoiled. Their commander, dashing forward, shouted orders to advance which went unheeded. "I am sorry to have to report to you," he yelled to Pakenham, "that the troops will not obey me. They will not follow me!" Pakenham spurred his horse to the front to stiffen the line. A shower of rifle balls from the rampart greeted his appearance. One shattered his right arm; another felled his horse. Pakenham quickly mounted again and rode after the retreating column, calling them to halt and reform their line. A hail of grapeshot enveloped Pakenham and his staff. The British general fell to the ground, was carried to the rear, and died within minutes. Seeing one officer after another fall before the American fire, the British soldiers rushed from the field in confusion, leaping over the dead that lay piled high on the ground.

Meanwhile, the column moving along the river bank executed their orders decisively and on schedule. Indeed, had they attacked, instead of feinting, they might have reached the rampart and scaled it. But their orders called for a demonstration only, so they held back. Then, when they saw what was happening to their comrades on the right flank they bravely crossed over to help them. It was a mistake. As they crossed, they were slaughtered by the death-dealing fire of the American sharpshooters.

On the west bank, despite the delay in getting there, the British easily outfought the undermanned and undergunned Americans. Morgan and Patterson only had time to spike their long-range guns before retiring. As the Americans fled in confusion, an adjutant ran after them crying, "Shame, shame! Boys, stand by your general." But they would not listen, they would not stop, and General Morgan dejectedly followed them on horseback.[12]

The battle ended at noon. The carnage among British soldiers was so frightful, said one observer, that it was terrible
to behold. The ground, he reported, was "covered with dead
and wounded laying in heaps, the field was completely red."
The shrieks and groans of the wounded, the convulsive and
sudden tossing of arms and legs "were horrible to see and
hear"; they shocked the Americans to silence and wiped the
smile of victory from their lips. "I never had so grand and
awful an idea of the resurrection as on that day," Jackson
said later. "After the smoke of the battle had cleared off
somewhat, I saw in the distance more than five hundred
Britons emerging from the heaps of their dead comrades, all
over the plain, rising up, and still more distinctly visible as
the field became clearer, coming forward and surrendering
as prisoners of war to our soldiers. They had fallen at our
first fire upon them, without having received so much as a
scratch, and lay prostrate, as if dead, until the close of the
action." Others were not so lucky, and many who were
wounded died a few days later. When the grim business of
counting the dead was done, the figures revealed that 2,057
British had perished, and—incredible though it seemed—
only 13 Americans had died. Most of these 13 were members of the Negro regiment, "who were so anxious for glory
that they could not be prevented from advancing over our
breast works and exposing themselves. They fought like des-
peradoes. . . ."

It was an unbelievable victory. Never had American arms
sustained so overwhelming, so complete a victory as this.
Never by force of arms had Americans proved so convinc-
ingly their right to independence. The victory was shaped in
part by British failure to coordinate their attack, Jackson's
success in massing tremendous fire power behind the rampart, and the excellent marksmanship of the American soldiers. Jackson had luck too, the luck that stalled the column
headed for the western side of the river.

For the next ten days, both armies stood poised, watching one another, not fighting except to lob cannon balls back and forth on occasion. The dead were buried, though Pakenham's body was encased in a hogshead of rum and shipped to Havana, Cuba. On January 19, what remained of the British force slipped away from their lines, retired to their ships and later sailed away. The major portion of the American force withdrew from its position on January 21 and entered the city of New Orleans amid scenes of unrestrained joy and happiness. The cathedral peeled forth a *Te Deum,* and in the square a triumphal arch was erected, supported by six columns, under which Jackson rode like a Roman Emperor.[13]

The day of the ceremony the streets teemed with people. The balconies and rooftops came alive with them. Everywhere there were people—in the great square opposite the cathedral, in the streets leading to the square, along the river bank—everywhere. They had come to honor their savior. Young ladies were dressed to represent Liberty and Justice. Other girls, dressed in white, wearing a silver star on their foreheads, and carrying baskets of flowers, appeared as States and Territories of the Union. Under the triumphal arch erected in the middle of the great square appeared two children standing on a pedestal holding a laurel crown. Then, into the square strode the Hero of the Battle of New Orleans, his head uncovered; an enormous cheer burst from the crowd. The people called and waved to him, their faces sparkling with joy. The General was gestured toward the cathedral which meant passing under the arch and between the rows of ladies representing the States. As he walked through the arch, the laurel crown was lowered to his head, and the ladies pelted the ground in front of him with flowers. One girl, representing Louisiana, stepped forward and congratulated the Hero on his victory in the name of the people of her state. At the cathedral porch the

Abbé Dubourg, resplendent in his ecclesiastical robes, saluted Jackson, and conducted him to his seat inside the church, while the organ and choir boomed the joyful noise.

In Washington, the news of the victory dissolved the pall that had enveloped the city since its invasion by the British. For months, it had lain buried in its shame and disgrace. The Congress was its usual factious and ill-tempered self. Moreover, representatives from the Hartford Convention were said to be headed for the capital carrying a threat of disunion if their proposals for constitutional revisions were not adopted. To add to the general gloom, a severe snowstorm struck the city on January 23, tying up roads for several days in every direction. Then, through the snow and the gloom and the general malaise came word of the stupendous victory at New Orleans, and Washington went "wild with delight." The city was illuminated. Crowds surged to the executive mansion, cheering the President, after which they moved to the houses of the Secretaries and the leading advocates of the war, saluting them with shouts of praise and approval. Newspapers broke out their largest type to headline this "INCREDIBLE VICTORY." Two lines from Shakespeare's *Henry VI* were reprinted in newspapers across the nation:

> Advance our waving colors on the walls
> Rescued is *Orleans* from the English wolves.

"Glory be to God that the barbarians have been defeated," said *Niles' Weekly Register,* "and that at *Orleans* the intended plunderers have found their grave! Glory to Jackson . . . Glory to the militia. . . . Sons of freedom—saviors of Orleans—benefactors of your country and avengers of its wrongs, all hail! Hail glorious people—worthy, thrice worthy, to enjoy the blessings which heaven in bounteous profusion has heaped on your country!"

A resounding refrain went up in all the states: *"Who would not be an American?"* To which was answered, *"Long Live the Republic."*

Hard on the heels of this splendid news came the announcement of the peace treaty signed at Ghent, Belgium, ending the war. Men rushed through the streets crying, "Peace! Peace! Peace!" And in Washington, the delegates from the Hartford Convention slipped quietly out of town before anyone could remember their mission.

Jackson was voted the thanks of the Congress along with a gold medal. He was now a national hero, and a hero of such dimensions that he soon became a legend in his own time. Secretary of War James Monroe sent him the congratulations of the President. "History records no example," he wrote, "of so glorious a victory, obtained, with so little bloodshed on the part of the victorious." State legislatures passed resolutions glowing with extravagant praise and acclamation. The New York resolutions, written by Martin Van Buren, called the victory at New Orleans "an event surpassing the most heroic and wonderful achievements which adorn the annals of mankind." Other states imitated the exaggeration, and in Europe, Henry Clay of Kentucky, one of the U. S. Commissioners who drew up the peace treaty, smiled with satisfaction over the New Orleans triumph. "Now," he said, "I can go to England without mortification." Indeed, now the whole country could hold up its head before Great Britain without mortification—thanks to General Jackson.[14]

CHAPTER IV

Florida

THE HERO LEFT NEW ORLEANS ON APRIL 6, 1815. AS SOON AS he received notification of the signing of the peace treaty, he lifted martial law, packed his things and returned to Tennessee. Many in New Orleans were delighted to see the old ramrod depart. While they were grateful to him for his victory, they resented his Draconian application of the law, his cavalier disrespect for their Creole temperament. Among many other irritations was Jackson's retention of martial law until the middle of March; as a consequence the Louisiana Senate purposely omitted his name from a list of officers to whom it extended its appreciation for saving the city. One incident that particularly rankled was Jackson's arrest of a federal district judge, Dominick A. Hall, for issuing a writ of habeas corpus freeing a legislator named Louis Louaillier, who had been jailed for writing a newspaper article that defied the city's military authority. It was bad enough to imprison Louaillier; worse to arrest a district judge; but to insinuate on the basis of trifling evidence that Hall was an agent of the English, as Jackson did, was vicious and dishonorable. Later, when Hall was released and martial law lifted, the judge summoned the General into court, held him in contempt, and fined him $1,000. Though respectful of Jackson's service, Hall said it was simply a matter of whether the law yielded to the General or the General yielded to the law, and as long as he sat on the bench Hall declared there was no question which came first.

Outraged by the judgment, even to the point of refusing

to accept a thousand dollars raised in a popular subscription by his friends, the Hero paid the fine. He did not defy Hall, much as he wanted to, because he was not prepared to accept the consequences of such a rash action. Also, he did not wish to blemish his victory with a quarrel he was certain to lose. The fine was remitted to him thirty years later by a grateful Congress and it gave him immense pleasure to accept it—both principal and interest.

With the war at an end, the United States Army was reorganized in the spring of 1815 into northern and southern divisions, each commanded by a major-general: Jackson for the southern and Jacob Brown for the northern. This was a pleasant arrangement and allowed Jackson to name the Hermitage as his headquarters, thus permitting him the luxury of attending his military business while managing his plantation and generally enjoying the life of a country gentleman and first member of his society. Occasionally he journeyed to Virginia and Washington on private or public assignments. Once, he stopped at Lynchburg, Virginia, where the whole population turned out to do him homage. There was a magnificent reception and banquet, and among the guests was Thomas Jefferson, now seventy-two years of age, who had ridden to Lynchburg from his home at Monticello to join in the tribute to the Hero. When asked to give a toast, Jefferson faced Jackson and said, "Honor and gratitude to those who have filled the measure of their country's honor." The General responded by toasting James Monroe, Secretary of War at the time of the Battle of New Orleans, a Virginian and Jefferson's friend. It was a gracious response, and the people thought it properly modest, though it may have masked an old wound that still festered. In other towns, there were more dinners and toasts, to which Jackson invariably replied by mentioning the patriotism and services of some local figure. He became so proficient at acknowledging the contributions of others and so unassuming in referring to his own "humble exertions" that many thought they caught the faint rumblings of a politician in search of popular support. Soon, Jackson was regularly touring the country—as he did after all his triumphs—letting thousands see their savior, letting them glimpse this impossibly gallant and brave defender of the Republic.

It was on one of these tours that the General heard disquieting news about the southern Indians: rumors that the Seminoles were hunting in Georgia in deliberate violation of the Treaty of Fort Jackson. Most of the Seminole settlements lay in Florida, except for Fowltown, which was situated in southern Georgia. But what really bothered Jackson and most Americans was the information that Florida was becoming a staging area for Indian raids on white settlements in the United States. Also, it was attracting runaway slaves from the entire southern region; it was a haven for them, a place to repair to once they escaped their masters. In fact a band of fugitive slaves from South Carolina and Georgia had already seized a fort on the Apalachicola River in northern Florida and encouraged other slaves to join them. Since this "Negro Fort" menaced all slaveowners it had to be obliterated. General Edward Gaines dispatched a force to the Apalachicola, blew up the fort, and killed 270 Negro men, women, and children.

That removed the Negro menace. The Indian problem remained, and by 1817, it had become more acute because of the increased numbers of white settlers and speculators who had moved into the Florida area and demanded the removal of the Indians. Among the speculators operating near Pensacola was a group of Tennesseans headed by Captain John Donelson and Major John H. Eaton, both of whom were urged to invest their money in the Pensacola project by General Jackson. Jackson believed Pensacola was destined for eventual annexation by the United States and then to develop into a rival of New Orleans for the Gulf trade.[1]

As the whites continued to penetrate the coastal region, there were repeated instances of forceable Indian removal from their lands. In revenge, the red men attacked white settlements and then fled to Florida for safety. President James Monroe sought to resolve the problem by purchasing Florida from Spain, a project long the ambition of the last three Virginia Presidents. When he failed in his quest, a series of filibustering expeditions were launched against the sanctuary to test the strength of the Spanish and the Indians. Both proved invitingly weak.

Then, in 1817, an elderly Scottish trader by the name of Alexander Arbuthnot arrived in Florida. He had come to trade knives, guns, powder, and blankets with the Indians in return for furs, beeswax, and corn. Arbuthnot was one of those rare white men who genuinely sympathized with the Indians and tried to help them. True, he also wanted their trade, but he seems to have treated the Indians decently and wrote letters to influential men pleading their cause. Later in 1817, a second Britisher arrived in Florida, one Robert Ambrister. A former lieutenant of the Royal Marines, this swaggerstick strutted around in his dress uniform, barked commands, and, in all, staged a performance the savages found delightful to watch. He, too, wished to aid the Indian cause, but unlike Arbuthnot he urged the Seminoles to fight the Americans, to stand up bravely and whoop their defiance.

Matters came to a head when one of the Indian chiefs seized land near Fowltown and announced he would resist any effort to oust him. General Gaines acknowledged this declaration of war by personally commanding an expedition that burned the town and killed its warriors and women. Nine days later, on November 30, the Seminoles took revenge by ambushing a large open boat as it floated down the Apalachicola carrying forty U.S. soldiers, seven of their wives, and four children. They scalped the adults and murdered the children by slamming them against the sides of the boat. Four men escaped by jumping overboard and swimming to shore. One woman was taken captive.

The fate of the Seminoles was now sealed. Their chastisement was a matter of necessity, and there was only one man who could inflict it with sufficient pain to teach them a lesson they would never forget. Although Jackson had been in slight disfavor with the administration because of his insistence that War Department orders not be transmitted to his subordinates except through him, Monroe asked him to forget their difference of opinion and work together to rebuke the Indian outrage against "our national character." On December 26, 1817, Secretary of War John C. Calhoun of South Carolina, ordered the General to Georgia to take command of the war. When the Indians heard they had been turned over to the tender mercies of Andrew Jackson they fled in terror into Spanish territory. To the red men all Jackson needed to do was simply point at them and they perished where they stood.

Previously the General had received a copy of General Gaines's order authorizing him to pursue the Indians across the border and "attack them within its limits . . . unless they should shelter themselves under a Spanish post. In the last event, you will notify this Department." The order was purposely vague and gave the widest latitude in coping with both the Indian and Spanish problems since obviously they were connected. Jackson rightfully presumed the order to Gaines now applied to him, but instead of impetuously plunging across the border as the administration expected him to do, the Hero cautiously wrote the President and requested permission to seize Florida. Disregarding his immediate superior, the Secretary of War, Jackson asked Monroe to come right out and acknowledge his intentions. Worse, he suggested a subterfuge. He told the President to signify through any channel, say John Rhea, a congressman from Tennessee, "that the possession of the Floridas would be desirable to the United States, and in sixty days it will be accomplished." Monroe later claimed he was ill when the letter reached him and that he never spoke to Rhea. Jackson, on the other hand, insisted that in February, 1818, he received the assurance he wanted and that he later burned the letter at Rhea's request. The General was wrong. There never was such a letter. Still, it is easy enough to understand what happened because Monroe's desire for him to take Florida was so clear at the time that Jackson later concluded permission must have been obtained and therefore that it came through Rhea. However, when his actions in Florida generated an international storm and the General was threatened with reprimand, Rhea said to him, "I will for one support your conduct, believing as far as I have read that you have acted for public good." Rhea would never have spoken this way had Monroe indeed employed him to authorize the invasion.[2]

Even so, Jackson did have a kind of permission, and it came directly from Monroe. When the President originally wrote the General patching up their small disagreement about lines of command he concluded his letter by implying very broadly that he desired the invasion of Florida. "This days mail will convey to you an order," he said, "to repair to the command of the troops now acting against the Seminoles, a tribe which has long violated our rights, & insulted our national character. The mov'ment will bring you, on a theatre, when possibly you may have other services to perform depending on the conduct of the banditti at Amelia Island, and Galvestown." Now, what was Jackson supposed to think those "other services" were? The letter went on. "This is not a time for you to think of repose. Great interests are at issue, and until our course is carried through triumphantly & every species of danger to which it is exposed is settled on the most solid foundation, you ought not to withdraw your active support from it." What great interests? Killing Seminoles? Hardly. The words "every species of danger" surely indicated the Spanish in Florida.[3]

Monroe's letter was mailed at approximately the same time Jackson sent his Rhea letter. They crossed each other in transit. No wonder the General later presumed he had received permission to seize Florida, particularly when he also received a letter from Rhea on January 12 (actually written too early to constitute the authority he requested) stating, "I expected you would receive the letter you allude to, and it gives me pleasure to know you have it, for I was certain it would be satisfactory to you. You see by it the sentiments of the President respecting you are the same." In addition, if the administration truly intended to respect the territorial integrity of Florida, the Hero was certainly not the man to send after the Indians. Monroe knew this; he also knew the General's hatred for the Spanish; he knew his intense nationalism. Indeed, he counted on that nationalism when he spoke of "other services" and mentioned that this was not the time to "repose" until "every species of danger" had been settled. The President's intentions were absolutely clear, only Monroe did not have the courage later on to take full responsibility for the events he had set in motion.

Additional evidence that Jackson had full authority for the invasion comes from the remark of William H. Crawford, Secretary of the Treasury, who told a friend that he changed his mind a short time later about the correctness of the General's conduct. "I have seen a letter," said Crawford, "shown me by the President that has perfectly satisfied me that General Jackson was fully justified in the course he adopted in the prosecution of the Seminole War in every respect." Crawford went on to say that if anyone tried to make trouble on account of the invasion, Jackson could probably use the letter to blow them all sky high.

In any event, the Hero had been ordered to subdue the

Indians, that much was certain, and he did it the only way he knew how: pursue them, kill them and burn their villages. Since the "banditti" had escaped to Florida, he gave chase, disregarding the prohibitions of international law and the restrictions of boundary lines. Commanding two thousand men from Fort Scott, he stormed across the frontier to St. Marks (where the Indians reportedly had gone), seized the fort and, to add insult to injury, lowered the Spanish flag and raised the stars and stripes. Jackson also discovered Arbuthnot at the fort—that "noted Scotch villain" as he called the kindly old man—and held him for trial. But learning that a body of Seminoles had assembled at the village of Chief Bowlegs (or Bolick as he signed himself) on the Suwanee River, the General quickly left St. Marks and scurried across a hundred miles of swamp, expecting to surprise the Indians and finish them off before they knew he was in the neighborhood. No such luck. When he arrived at their village, the Seminoles were gone. They had been forewarned by a letter sent by Arbuthnot to Bowlegs and had disappeared into the swamps. Soon afterward, the swaggering Ambrister blundered into Bowlegs' camp, not knowing his friends had departed and that the village was now occupied by American forces. Both he and Arbuthnot were tried for inciting the Indians to war against the United States and furnishing them with arms and provisions. The court found both men guilty and sentenced Arbuthnot to be hanged and Ambrister to be shot. However, when one of the members of the court asked for a reconsideration of Ambrister's sentence since the former British marine was obviously an adventurer and had acted without malice, the court changed its verdict and sentenced him to fifty lashes on the bare back and a year's imprisonment at hard labor. But Jackson had no mercy for such "criminals." He reimposed the original sentence and ordered Ambrister "to be shot to *death*" and Arbuthnot to be "suspended by the neck with a rope until he is *dead*." On April 29, 1818, the sentences were carried out. As Arbuthnot was jerked up the yardarm of his own ship a group of Indians stared at him, scarcely understanding how so powerful a friend could be summarily executed by a foreign intruder. With his dying breath, Arbuthnot swore his country would avenge his death, but the Indians found his words less than convincing as they watched his body sway in the Florida breeze.[4]

For Jackson, the executions were a dirty but necessary business. He did not stay to witness them. Instead, he garrisoned Fort Marks with two hundred troops and set out for Fort Gadsden, intending to scour the country west of the Apalachicola and clean out the remaining hostile Indians. On route, he discovered only a few Seminoles with any fight left in them, and these were easily subdued. Thus, for all intents and purposes, the war with the Indians was over.

Turning then to the Spanish, Jackson thrust his army at the "hated dons," seizing Pensacola on May 24 and sending the governor fleeing for the Barrancas. Matching Spanish arrogance with his own, the Hero flatly informed the governor that he was assuming control of the country "until the transaction can be amicably adjusted by the two governments." Obviously he presumed the United States government would legitimatize his seizure since the President had virtually ordered him into the country in the first place. But the Spanish governor thought otherwise and demanded the General's immediate removal from Florida, whereupon Jackson captured Barrancas and had the governor, his soldiers, and civil officials escorted to Havana. The simple violation of the frontier had now been compounded to include the destruction and removal of Spanish authority from Florida. Still it could have been worse, for Jackson, speaking of the governor, later remarked, ". . . All I regret is that I had not . . . hung him. . . ."

With the Spanish ousted from Florida, with the Indians

subdued—in short, with the job of eliminating "every species of danger" remarkably well done—the General stationed a garrison at Pensacola and returned home to Nashville. He was greeted with the familiar demonstrations: parades, speeches, and public dinners. If possible his stature as hero was even greater than before, and the American people acclaimed him as the glory of the nation. But in Washington, this glory seemed a trifle too much. The President, the Cabinet, and the Congress faced not simply a conquering hero, but one who had committed all manner of prodigalities, including the invasion of Spanish territory, the annihilation of Spanish rule in Florida, and the execution of two British subjects—all of which was bound to create a spectacular international commotion.

When official news of these extraordinary exploits reached Washington, the Spanish minister, Luis de Onis, splattered the administration with indignant protests. Unnerved by Jackson's excesses and Onis' protests, the "fickle-minded" Monroe hesitated to protect his commander and instead summoned his Cabinet for an opinion. They gave it to him. Unless he were prepared for an international blowup with both England and Spain, they said, he had better disavow the Florida invasion and censure Jackson for exceeding his instructions. Calhoun, indignant because the Hero had gone over his head to Monroe for authorization to seize the territory, voted with the majority. William H. Crawford, the Secretary of the Treasury, who expected the presidential nomination in 1824 and feared the General as a rival for popular votes, also approved the censure but changed his mind some time later when Monroe showed him a copy of his original letter to Jackson. Only the dour, tough-minded, and brilliant Secretary of State, John Quincy Adams of Massachusetts, defended Jackson, arguing that the General's action had been defensive and required by his principal duty of subduing the Seminoles. When the

Cabinet discussion ended, the President was forced with the dilemma of approving Jackson's actions, which constituted an undeclared act of war against Spain; or disavowing them, and running headlong into the shock of the Hero's popularity, as well as giving the appearance of truckling to Spain. Monroe had anticipated some trouble with Spain but not with the American people. Under the circumstances, therefore, he wisely decided to support Jackson, even though he could not publicly admit his complicity in the invasion. When the time came to hoodwink the Spanish about how all this had come about, Adams blandly told them that Jackson had not been ordered into Florida but was forced to enter the country to prosecute the war against the Seminoles. The Secretary added that Pensacola and St. Marks would be restored, but he advised Spain either "to place a force in Florida adequate at once to the protection of her territory" or cede the country to the United States.[5]

Jackson—rightly convinced that his actions were consistent with the true wishes of the administration—was shocked by the government's dilemma. He immediately reduced the problem to one of personalities, suspecting some enemy at work to tarnish the luster of his victory. He did not have far to look. There he was, big and hulking, sitting in the chair of the Secretary of the Treasury: none other than William H. Crawford. Jackson's dislike of Crawford went back many years and originated over the disposal of lands taken from the Indians by the Treaty of Fort Jackson. The Cherokees claimed that the General had illegally annulled their rights to certain lands in the Gulf area, and they successfully argued their case before Crawford, who returned four million acres to them. Jackson was furious over the decision. He had spoken to Crawford and had urged him to refuse the claim, but the Secretary paid him no heed. Raging over the violation of "his" treaty, the General nailed Crawford as a jealous rival, intent upon diminishing his reputation and fame.

With pressure mounting in the capital for his censure, Old Hickory darted across the mountains and entered Washington on January 27, 1819. Not a minute too soon. Those in Congress who opposed the administration seized on the invasion as the perfect instrument with which to bludgeon Monroe and his Cabinet. Henry Clay led the pack. His enmity toward the administration was rooted in his ambition. He had wanted to be Monroe's Secretary of State because that post supposedly carried the succession to the presidency, and when he was offered the War Department instead, he indignantly refused it and so became the leader of the antiadministration forces in Congress. In the House of Representatives Clay rose to do battle. His face gaunt, his voice piercing, he denied personal animosity toward Jackson or toward the administration. He was simply acting from principle, he said, just as he always did. The Representatives "may bear down all opposition . . . even vote the general the public thanks," he said; "they may carry him triumphantly through this house. But, if they do, in my humble judgment, it will be a triumph of the principle of insubordination—a triumph of the military over the civil authority—a triumph over the powers of this house—a triumph over the constitution of the land." He prayerfully ended by hoping it might not also "prove . . . a triumph over the liberties of the people."

Jackson's response to this assault was instant and abrasive. "The hypocracy & baseness of Clay," he wrote Major Lewis, "in pretending friendship to me, & endeavouring to crush the executive through me, make me despise the Villain."

But more important than the General's usual and predictable eruption when abused was the political skill with which he now met the attack. He sent copies of Clay's speech to his friends in the West and asked them to have it published in the newspapers. "I hope the western people will appreciate his conduct accordingly," he wrote rather pointedly. "You will see him skinned here, & I hope you will roast him in the West." In addition, because John Quincy Adams had defended him so ably, he asked that the newspapers accord the Secretary "a proper ulogium." Also the President. "I knew I could not be mistaken in Mr Munroe firmness, he must be supported." As for Calhoun, that brooding man with dark eyes set in a glowering face, the Hero wrote that he had received assurances of his loyalty. Calhoun "has professed to be my friend, approves my conduct and that of the President." Thus, although Jackson did not know it at the time, the hypocrisy of Clay was matched by the hypocrisy of Calhoun—but with a difference. Calhoun kept his secret.[6]

Not only did Jackson campaign in his own defense through friends in the West but also through his partisans in Congress; and it paid off, for when the House voted on the resolutions disapproving the administration's actions in Florida, the members voted them down. With that, the Hero claimed vindication. Then a few days later—no doubt acting from principle again—Clay visited Jackson at his hotel to assure him there had been nothing personal in his effort to cut him down in the House. Fortunately, the General was out when Clay called and so was spared the unpleasantness of a face to face encounter.

The final disposition of Florida was neatly handled by Adams. After considerable negotiations, Spain recognized that in view of her losses in South America, and the ease with which the Americans could retake Florida, it made sense to sell the troublesome territory. Accordingly, the Adams-Onis Treaty was signed on February 22, 1819 by which Florida was sold for five million dollars, and the western boundary of the Louisiana Territory was fixed at the Sabine River.

Only one problem remained: the diplomatic disposal of Ambrister and Arbuthnot. Naturally Britain assumed an indignant stand over the executions, although, as Adams guessed from the very beginning, she had no intention of going to war over the incident. Richard Rush, the United States minister to Great Britain, and Adams argued that both Ambrister and Arbuthnot were partially responsible for inciting the Indians to war against American frontiersmen. The evidence was persuasive, so the British chose not to contest the severity of the punishment or to ask an indemnity.

Jackson glowed. Now that his laurels were adjusted to his satisfaction and his honors and victories all intact, he left Washington and began a triumphal tour of the Middle Atlantic states, visiting the leading cities of Baltimore, Philadelphia, and New York and seeking a popularity to equal that which he already enjoyed in the West and along the frontier. Everywhere his presence prompted celebrations, parades and illuminations, and the General responded instinctively by waving to the crowds and tipping his hat. A proud man might find tipping his hat impossible; but a politician can spring it off his head at the first sight of a crowd, and Jackson was one of the fastest hat-springers in the business. At a banquet in Philadelphia, after hearing himself toasted for his gallant services to the nation, the Hero replied by raising his glass to, "The Memory of Benjamin Franklin." The mob roared its delight. In New York, at a Tammany Hall celebration, the General made a slip by saluting the builder of the Erie Canal, De Witt Clinton, not knowing the Tammany men detested Clinton. But it was a mistake any outsider might make, so the Tammany men just laughed it off. Clinton, however, was ecstatic over the toast, and as long as he lived, though mercurial in almost every other respect, he remained loyal and devoted to Andrew Jackson.

After basking in the warm affection of eastern Americans and sporting himself in the leading seaport cities, Old Hickory returned to Tennessee. But he did not cease his politicking. He surrounded himself with dedicated friends, old comrades-in-arms and neighbors and assigned them various tasks to advance his ambitions. This group was the Nashville Junto, a corps of extraordinarily gifted politicians whose principal purpose became the election of Andrew Jackson to the presidency of the United States. Among them were John H. Eaton, an able penman who articulated Jackson's views in print; John Overton, a planter, bank president and one of the General's oldest friends; Felix Grundy, a rough-and-tumble politician of great sagacity and shrewdness; and William B. Lewis, a tall, powerful-looking man of limitless industry and dedication who was Jackson's "home confederate." A few of these men gravitated to Jackson because they hoped to use his popularity to achieve local political advantages. Others felt an intense personal loyalty. But, whatever their motives they developed into a skillful and inventive political clique that effectively managed the Hero's presidential campaign.[7]

It was because of this ambition that Jackson overcame an initial hesitancy and agreed to accept the territorial governorship of Florida when Monroe offered it to him in 1821. Another reason was his desire to savor the delight of presiding over the formal ouster of Spanish rule from Florida. Also, it served as a kind of vindication for the attempted censure in the House of Representatives. Consequently, he accepted Monroe's offer and was commissioned in March, 1821. In addition to receiving the territory from the Spanish, he was also permitted to exercise all the authority previously held by the captain-general of Cuba and his two governors of East and West Florida until the next session of Congress. He was authorized to suspend officials not appointed by the President, but he was forbidden to lay new taxes or grant public lands.

Upon accepting the office, Jackson resigned from the army. His illustrious military career now concluded, he was about to embark on a new career in search of even greater fame. Together with his wife and the "two Andrews," as the General called his adopted son, Andrew Jackson, Jr., and his nephew, Andrew Jackson Donelson, he left the Hermitage in the spring of 1821 and traveled to Florida by way of New Orleans. Rachel, grown stout, dark complexioned and exceedingly pious, was aghast by what she saw along the route, especially in New Orleans. "Great Babylon is come up before me," she exclaimed. "Oh, the wickedness, the idolatry of this place! unspeakable the riches and splendor." She wept over the city, just as any Christian lady might be expected to do, finding peace of mind only in Holy Scripture, especially Psalm 137. When Rachel left this "Babylon-on-the-Mississippi," she discovered the wickedness repeated in other cities—everyone she visited, without exception. No one respected the Sabbath, she complained; no one had the true religion. "I know I never was so tried before, tempted, proved in all things," she wrote. "I know that my Redeemer liveth, and that I am His by convenant promised."

Taking this "saint" into the swamps of Florida tried not only her soul but her digestive tract as well. Neither she nor Jackson cared particularly for the food or the climate, and they arrived in the territory just as the hottest season began. And there were other discomforts. Colonel José Callava, the Spanish governor, treated them contemptuously when they arrived in Pensacola. He steadfastly refused to sit down and work out details for the transferal until he received explicit orders from the governor of Cuba. Incendiary words shot back and forth before the difficulties were finally smoothed over and the territory formally ceded to the United States on July 17, 1821.

On the appointed day, American troops were massed in the public square opposite the Spanish guard. Jackson and his aides passed between a double line formed by Spanish and American troops and entered the Government House. After the formal transfer took place, the Spaniards marched to their ships and sailed to Cuba. The American flag was run up the staff; the artillery fired salutes; and the regimental band played the "Star Spangled Banner."

Colonel Callava remained as a diplomatic agent finishing up the king's business. And he nearly drove Jackson out of his mind. First, there was an argument over the ownership of cannons in the forts; next, they quarreled over provisions for the Spanish garrisons during their journey from Florida to Cuba; then, there was a dispute about documents involving the efforts of certain wealthy men to defraud the mulatto children of a deceased landowner. On behalf of the children, the General demanded the documents forthwith; Callava refused, first on one pretext, then another. When Jackson could stand this stalling no longer, he ordered Callava to prison. "I did believe," he wrote, "and ever will believe, that just laws can make no distinction of privilege between the rich and the poor. And that, when men of high standing attempt to trample upon the rights of the weak they are the fittest objects for example and punishment."[8]

Eventually Callava was released. And he no sooner left prison than he shot up to Washington as fast as he could get there to protest his treatment to the Secretary of State. Again, Adams defended Jackson and blamed the unpleasantness on Spanish procrastination in processing the transfer and concluding business. But Adams erred; there was wrong on both sides. Jackson had not treated Callava with the respect due him as a diplomatic agent. He was his usual highhanded self, and he tended to respond to situations from a deep-seated prejudice against the Spanish, expecting nothing from them but treachery and deceit. Although justice was on Jackson's side in the dispute over the documents, his over-all conduct toward Callava was arbitrary and improper.

Jackson's mistreatment of Callava was not unique, however. Anyone who got in his way felt the back of his hand. As territorial governor, he frequently presumed absolute authority, which weekly brought howls of angry protest. Soon, complaints to Washington were a common occurrence, and John Quincy Adams remarked that he dreaded the arrival of the mail from Florida because he never knew what new incident the General had contrived that would necessitate a painful session with the Spanish minister.

Nevertheless, on the assumption that Florida needed a government that was simple and energetic, Jackson provided the necessary machinery. He set up a municipal government for Pensacola with one of his pensioners as mayor or alcalde. He issued ordinances designed to safeguard the public health; and he established counties and county courts. The organization of Florida went forward with such dispatch that, although Congress later annulled some of his acts, the initial frame of government proved to be practical and worthwhile.

But almost before the summer of 1821 ended Jackson was sick of Florida, its problems, and its climate. He was fifty-four and suffering from chronic diarrhea and indigestion. He harbored suspicions that the swamps were meant to sink his political ambitions. Moreover, he was distressed by the failure of the administration to follow his recommendations in the distribution of patronage. Too many jobs were going to men of whom he disapproved—and over his written protests. Within a few months, he sensed he was employed in an inferior role, his judgment subject to higher and critical scrutiny. Finally he felt Monroe had lost confidence in him as a result of the Callava affair and would gladly accept his resignation. Frustrated, suspicious, ill, and a little resentful of Washington's unflattering attitude toward his tenure, he threw up his job and left Florida on the pretext that his wife's health was bad. After arriving at the Hermitage early in November, 1821, he formally resigned the governorship, which President Monroe accepted as of December 1.

Because Jackson was not the kind of man who could submit tamely to any treatment the government chose to award him—whether a "well done" or a crack over the knuckles—his subordinate role as governor of the Florida Territory was predictably short. Whenever he was given an assignment, Jackson demanded the fullest authority and the fullest governmental support. Without these, he could not operate. In riding away from his job he notified the President that, "I have determined to take a little respite from the laborious duties with which I have been surrounded and leave the charge of the Floridas to the Secretaries appointed for the same." Just like that. Without fear or worry and without giving it a second thought he abandoned the swamps of Florida and returned home. No doubt, the swamps of Washington were more to his liking.[9]

CHAPTER V

Quest for the Presidency

THERE ARE MANY REASONS TO EXPLAIN HOW AND WHY ANDREW Jackson became the President of the United States. But surely the most important was his own intense ambition, added to which were the political skills he developed over the years, and his fantastic popularity resulting from his victory at New Orleans. That popularity was fostered and nurtured by Jackson, and beginning shortly after the start of Monroe's second term as President, it began to manifest itself politically in the several states, usually taking the form of resolutions expressing support should he decide to strike for the executive mansion in Washington.

Grass-root sentiment alone can not make a president, however. It merely alerts politicians to available candidates. What produces presidents is organization, one structured by skillful operators of the political art who give the candidate the engine he needs to sustain the long trip to the White House. Such an engine, in Jackson's case, faced formidable obstacles because the presidency, since the founding of the nation under the Constitution, had always been reserved for men of demonstrated experience and ability in government, not military heroes, and certainly not Indian fighters.

But the country was rapidly changing in the 1820's. In every field of endeavor, a new generation was pushing forward, anxious to seize power, leadership, and status from an older political, economic, and social elite. These "men on the make" found in Andrew Jackson a symbol of their own ambitions and hopes. They discovered analogies between his career and achievements and their own strivings for advancement. He became a classic example of the self-made man who vaulted from log cabin to the White House all on his own; he was the personification of the American success story. Presumably if an orphan boy from the piney woods of Carolina could seek the highest office in the country, there was no reason why an ordinary citizen could not aspire to wealth and social status by relying on his own talents to get what he wanted.

Andrew Jackson was fortunate to stride across the national scene just as these changes worked their subtle effects upon the people. Soon politicians began to sense the response the General's popularity produced among the electorate, and how they might capitalize upon it in their states and direct it toward their own particular goals. By the beginning of the 1820's, there was a pronounced political shift toward Jackson that basically expressed the larger shifts in American life taking place all over the country.

Of equal importance was Jackson's own recognition of his appeal—an awareness of the identity people unconsciously felt toward him and his struggles. In the political battles ahead, when shaping strategy and seeking support, he instinctively turned to the great mass of the people, invited their approval, and reminded them that his fight was essentially theirs. "His strength lay with the masses," wrote one perceptive political observer, "and he knew it."

Naturally, it was in Tennessee where Jackson's popularity was first summoned to serve political democracy. The Nashville Junto of Overton, Eaton, Lewis, and Grundy had inherited the old Blount faction and were doing splendidly with it when the economic panic of 1819 struck the nation a crippling blow. In part this collapse was aggravated by the sudden contraction of loans and credit initiated by the action of the Second Bank of the United States. Because so many of the banks in Tennessee were owned and operated by the Junto, it suffered a beating at the polls in the next election by the John Sevier faction, now led by Colonel Andrew Erwin and Senator John Williams. To recoup their loss, the Junto turned to Jackson and asked him to run for office, even though some of the members were fearful of the Hero's antibank and anti-paper prejudices. The General, fresh from the disappointments of Florida, readily consented, and by the summer of 1822, after organizing the necessary support through newspapers and local meetings, the managers succeeded in inducing the legislature of Tennessee to nominate Andrew Jackson for the presidency of the United States.[1]

To strengthen Jackson's candidacy throughout the nation and disguise his somewhat threadbare credentials, the Junto resolved to run him for a seat in the United States Senate. This would not only provide instant-grooming for the presidency but his election would mean the defeat of John Williams, a leader of the Sevier faction who was then standing for re-election. A victory would also afford Jackson personal satisfaction, for Williams had supported the resolutions to censure him for the invasion of Florida. Still it was a risky business. Williams was not without powerful friends in the Tennessee legislature, and if he defeated Jackson he would rub out the Hero's presidential plans in a single stroke. The Junto braved the risk, however, and through careful attention to the details of politics squeezed out a victory in the legislature in October, 1823, by a vote of 35 to 25.

The Jackson candidacy for president, so propitiously inaugurated, found immediate response across the nation. Of special significance was the declaration of support from Pennsylvania, which originally leaned toward Calhoun but now switched to the Hero and nominated the South Carolinian for Vice-President instead. Other states in the North and South also registered their enthusiasm for the

Tennessean, and Jackson's campaign rushed off to an impressive start. Good thing, too. The field was already peopled with anxious candidates, all of them men with distinguished careers of public service, all men of experience, training, and reputation. First, there was William H. Crawford—"I would support the Devil first," growled Jackson—who had been campaigning since 1816. Second, there was John Quincy Adams of Massachusetts—"a man of first rate mind of any in america as a civilian and scholar," allowed the General—who could continue the tradition of distinction brought to the presidency by such luminaries as his father, John Adams, Washington, Jefferson, and Madison. Finally, there was that "base hypocrit," Henry Clay of Kentucky, who could match Jackson's Western background and ambition but not his popularity or sense of caution.

In the summer of 1823 it looked for a time as though the field would be reduced by one. Crawford suffered a stroke that left him temporarily speechless, sightless and immobile. But his Congressional friends, led by one of the smoothest operators in Washington, Martin Van Buren, the junior Senator from New York, would not withdraw him from the race. Rather, they summoned a Congressional caucus on February 14, 1824, and formally nominated him for the presidency despite the appearance at the meeting of only sixty-six men. Jackson, who had taken his seat in the upper house on December 5, 1823, and was eligible to attend the caucus, wisely elected to stay away. Had he put in an appearance, there would have been acute embarrassment all around.

It was now twenty-six years since Jackson first served in the Senate. In that time he had not perceptibly advanced in his understanding of the legislative processes. Yet, in the first session of Congress after his return in 1823, he demonstrated maturing powers as a politician, especially in his

conduct toward colleagues whose hands were close to the
levers of political power. For example, when he took his seat
in the Senate he found sprawled beside him his old adversary, Thomas Hart Benton, now the very capable Senator
from Missouri. Several friends offered to mediate their
quarrel, and although the rivals served together on the military committee and were more and more thrown together,
nothing came of it. But not many weeks passed before
Jackson took the full measure of Benton's legislative skill,
and he learned how powerful he was in Missouri. At the
moment Benton supported Crawford, though he was
unhappy about it after the latter's stroke. So Jackson—
whose reputation for undying enmity was vastly exaggerated where politics was concerned—stepped up to Benton
one day and asked about the health of his wife. Benton
quickly responded and returned the inquiry. A few days later
Jackson called at Benton's lodgings and, not finding him
home, left his card. After that, the first time the Missourian
saw the General he bowed. Jackson immediately shot out
his hand, the two men shook, and in that handclasp dissolved their old dispute—to their mutual advantage. Benton
was one of the first Crawford men to desert to Jackson, and
his influence in Missouri and in Congress later proved of
inestimable value to the Hero in winning elections and in
legislating the presidential program.

Old Hickory dissolved many other obstacles to his candidacy by the time the Congressional session ended. He may
not have played a significant role in legislative history, but
by his grace, dignity, and deportment he convinced many
Congressmen that he was an urbane and highly qualified
presidential candidate, not the rough savage from the back-
woods as many of his enemies contended. In fact the hyper-
critical Daniel Webster, Congressman from Massachusetts,
said he was the most presidential-looking of all the candidates before the American electorate, and, in this age, what

a candidate looked like was becoming increasingly more important.

A great many people around the country looked at Jackson just as Webster had, and in the fall election registered their approval, giving him 152,901 votes, as against 114,023 for Adams; 46,979 for Crawford; and 47,217 for Clay. But presidential elections are not determined by popular vote; it is the electoral vote that counts, and in 1824 none of the candidates received the necessary electoral majority. Jackson won a plurality of 99; followed by Adams with 84; Crawford with 41; and Clay with 37. In the vice-presidential contest John C. Calhoun was elected over token opposition.[2]

Since no one had a majority, the 12th Amendment to the Constitution directed the House of Representatives to select the president from the three men with the highest electoral votes. Each state would cast a single ballot, determined by its delegation in the House, and a majority of states was necessary for election. Because the 12th Amendment limited the candidates to the top three, Henry Clay was automatically excluded from the contest—which was unfortunate for him, because as Speaker of the House he exercised enormous influence over the state delegations and probably could have convinced them to make him President. Instead, he became the key man in selecting Monroe's successor. In choosing his candidate, he started off by rejecting Jackson almost out of hand because the General was his rival for Western votes as well as a "military chieftain" with insufficient governmental experience. Next, Crawford was dumped, because he was still physically incapacitated and because his extreme States' rights position clashed with Clay's "American System," an intensely nationalistic program of high tariffs, internal improvements, and a sound currency and banking system. That left Adams. And Clay, in selecting the hide-bound New Englander as his candidate, was impressed not only by Adams' very respectable qualifications for the office but also his commitment to a nationalistic program of economic and cultural development. However, to make certain Adams appreciated Clay's hopes for the future, a private conversation was held between the two men, at the conclusion of which it was understood that if Adams won the presidency, Clay would be his Secretary of State.[3]

But they were not the only politicians arranging an accommodation. Candidates and their managers scurried after every vote they thought was negotiable. Jackson was accused of seeing Crawford and offering him anything as the "price of his cooperation & support." At one point in the scramble James Buchanan, the Congressman from Pennsylvania, came to Jackson with a story that Clay's friends had indicated their readiness to "end the presidential election within the hour" if the General would first declare his intention of dismissing Adams from the State Department, and then, presumably, replacing him with Clay. Jackson was cautious in his reply, asserting he thought well enough of Adams but also that his good opinion did not necessarily mean he would retain him. What he was trying to do by this dodge was avoid commitments, steer "clear of the Society of both intriguers, & Caucus mongers" and yet keep himself attractive to the largest number of Congressmen. He posed as the disinterested statesman, above the political clamor. "If ever I fill that office," he once told Lewis, "it must be the free choice of the people—I can then say I am the President of the Nation—and my acts will comport with that character." So, all during the long winter preceding the House election, he stayed home with his wife, "chatting and smoking our pipe and thinking of our Tennessee friends." This made for a very homey scene. It also kept him out of the White House. In January, 1825, Clay announced his support of Adams and invited all the state delegations in the House to follow his lead. Naturally stories circulated that Clay had sold out to Adams in return for the State Department, but this Clay angrily denied and he expended so much energy in his denial that it was almost impossible to believe him.

The day of the House election, which was cold and snowy, came on February 9, 1825. A majority of thirteen states was necessary for a choice, and according to the advance calculations of the shrewdest politicians in the capital, the Adams-Clay coalition had twelve states; Jackson had seven; and Crawford, four. The strategy of the Crawford men, led by Martin Van Buren, required them to keep Adams from winning the thirteenth state and maintain a deadlock over a protracted period of time. That way, the delegations might be talked into switching to Crawford to break the deadlock. Obviously the key to the scheme lay in holding the thirteenth state in abeyance, and that state turned out to be Van Buren's own state of New York.

New York's delegation consisted of thirty-four men, split evenly between Adams and Crawford. Thus, the shift of a single delegate could swing New York one way or the other. Then, just as Van Buren's scheme seemed to be speeding along as planned, one of Crawford's pledged delegates, Stephen Van Rensselaer, was reported wavering. Alarmed, Van Buren spoke to him and finally persuaded him to renew his promise to stick to Crawford. But on the morning of the election, Van Rensselaer encountered Clay and Webster on his way to the House of Representatives. They hustled him into the Speaker's office where, in relay, they battered him with a range of arguments guaranteed to frighten him and convert him to Adams. Heroically the old man held out, but by the time they let him go he was badly shaken by their reminders of what his vote might mean in terms of his large estates in New York.[4]

Once in his seat back in the House, Van Rensselaer could not keep the doubts from crowding into his mind. Because he was an intensely religious man and wanted desperately to honor his pledge, he lowered his head to the edge of his desk and prayed for divine guidance. When the ballot box was brought to his place Van Rensselaer slowly raised himself, sighing over the absence of any heavenly murmur in his ear. Then, suddenly, he spotted an electoral ticket lying on the floor with the single name of "John Quincy Adams" written upon it. To Van Rensselaer, it was a sign, an answer to his prayer. Dutifully, he picked up the ticket and thrust it into the box. New York voted for Adams, and with thirteen states the New Englander was elected the sixth President of the United States.

Of the eleven states that had voted for Jackson in the electoral college, four—Louisiana, Maryland, Illinois, and North Carolina—deserted him in the House election. Crawford's three states—Virginia, Georgia, and Delaware—stood fast, and he picked up the additional state of North Carolina. Adams' majority included the six New England states, Maryland, Louisiana, New York, Illinois, Ohio, Missouri, and Kentucky.

Jackson took his defeat with dignity and grace. Yet it was a stinging rebuke in view of his popular and electoral vote. Still Adams was the President according to law, and the General had no choice but to accept it with magnanimity. At a reception given by James Monroe on the night of the election, Jackson walked up to the President-elect and extended his hand. "How do you do, Mr. Adams?" he said. "I hope you are well, sir." "Very well, sir," replied Adams. "I hope General Jackson is well."

That was Jackson at his most self-controlled—with maybe a little bitterness showing since he apparently did not congratulate Adams on his election. Then came the explosion. As soon as the President announced his intention of appointing Clay his Secretary of State, Jackson burst out in an impassioned rage. "So you see," he roared, "the *Judas* of the West has closed the contract and will receive the thirty pieces of silver." Convinced beyond argument that Clay and Adams had concluded a "corrupt bargain" to steal the presidency from him, the Hero poured out his wrath at them, cursing them for their villainy. Of course this public frenzy was carefully staged in order to create political capital, for the General's public outbursts were never impulsive or uncontrolled (and it did make extraordinarily effective propaganda); even so he sincerely believed that a monumental fraud had been perpetrated against the people and against himself. Predictably, the charge of a "corrupt bargain" became a rallying cry for a democratic crusade in 1828, at which time the vile corrupters, thieves, aristocrats, and other assorted enemies of "Jackson and the People" would be expelled from office.

Almost immediately, Jackson began to organize his opposition to the Adams administration. His first object in creating this opposition was John C. Calhoun. Even before the end of the Congressional session it was reported by Senator Rufus King that "a Party is forming itself here to oppose Mr. Adams' administration. . . . I understand that a Dinner takes place today [at] the Quarters of [the South Carolina] Delegation, when Gen'l Jackson, Mr. Calhoun . . . and others are to be guests. . . . This first step may serve to combine the malcontents." Some of the General's advisers, notably Lewis and Sam Houston, tried to warn Jackson that Calhoun was a hypocrite who had urged his censure over the Florida affair, but the foxy Hero, who appreciated the political strength of the new Vice-President, disregarded the warning and went right ahead with his plans for an alliance. For his part, Calhoun did not know which way to turn. He had been overwhelmingly elected Vice-President, thanks to the backing of both the Adams and Jackson electors. However, when Clay was elevated to the State Department and seemed to have stolen a march on the presidency, Calhoun's vacillation disappeared, and he threw his support to Jackson. A

Washington newspaper, the United States *Telegraph*, was established to serve the ends of the new alliance, and Calhoun's close friend, Duff Green, was invited from St. Louis to edit it.

Meanwhile, Jackson returned home to prepare for the 1828 election. The Tennessee legislature obligingly renominated him in October, 1825, and for the next three years he relentlessly toiled to recapture what the "thieves" in Washington had stolen from him. He commenced a well-organized, well-financed, and well-directed campaign for the presidency. He had learned in 1825 that, without the constant application of the arcane art of politics, popularity meant little, at least not while sinister men lurked in the halls of Congress. Jackson did not need to learn the lesson twice. He resigned from the Senate after only two sessions and supervised the establishment and direction of a Central Committee in Nashville, which corresponded with similar Jackson committees around the country. Newspapers were founded by the dozens in many of the northern states to launch the Jacksonian crusade, and in Washington a cadre of Congressmen committed to the Hero started holding regular caucus sessions to map strategy for defeating the Adams-Clay coalition. Soon, a wide channel of communication between Nashville and Washington was opened through an exchange of letters among Eaton, Houston, Benton, Jackson, Lewis, Alfred Balch, and many others.[5]

Late in 1826, an even more important alliance was concluded among the antiadministration factions, when Martin Van Buren, the short, fair-complexioned "Red Fox," or "Little Magician," as he was called, decided to jump on the Jackson bandwagon—since it was obviously going places—and bring his Crawford party with him. The Hero, despite his contempt for Crawford personally, made the transition as painless as possible for his new recruits by telling them how much he agreed with them about States' rights. Not, however, that he opposed "domestic manufactures and internal works" as a general policy, since that would frighten the more nationalistic-minded of his following. He thought the Crawford men would be happy to know he was extremely sympathetic, if not totally committed, to their point of view.

As the Jackson-Calhoun-Van Buren combination took shape, politicians of all creeds, sections, and interests began to orbit around the new political axis. Displaying ever improving qualities of political skill, Jackson, in acknowledging these disparate groups as members of his party, exercised extreme caution in defining his position on current issues. For example, he did not advertise his prejudice against banks and paper money; as for the tariff, he shrewdly placed himself on the side of a "careful Tariff," one that occupied a "middle and just course"—wherever that was. And, although he admitted opposition to federally sponsored public works, he softened that straightforward expression of opinion by announcing his approval of government distribution of surplus revenues to the states to permit them to undertake their own improvements. On one point, however, he was absolutely clear. As President, he expected to "purify the Departments" and "reform the Government." He told Amos Kendall, one of his staunchest supporters in Kentucky and editor of the *Argus of Western America*, that he would remove all men from office "who are known to have interfered in the election as committeemen, electioneers or otherwise. . . ." Also slated for the ax were "all men who have been appointed from political considerations or against the will of the people, and all who are incompetent." During the campaign the demand for reform was expertly handled by newspaper editors, who convinced the masses that Jackson's political head-hunting was really a crusade to purge the national government of corruption and privilege.

As the campaign picked up momentum, the party emerging around Jackson came to be called the Democratic party. The General's role in the creation of this remarkable organization was of primary importance. In political science, he was not profound or particularly knowledgeable, but his instincts were superb, and he knew how to encourage popular trust and faith in himself and in his party. But the campaign itself unfortunately degenerated into a mudslinging contest of incredible proportions. The candidates and their wives were all accused of immoral conduct. Jackson's marriage was twisted into the most loathsome type of propaganda. Charles Hammond, editor of the Cincinnati *Gazette*, was the chief propagator of this filth, and because a great deal of it was printed in the *National Journal*, an administration newspaper in Washington, Jackson blamed both Adams and Clay for instigating the scurrilous reports. Although the President had nothing to do with Hammond, he did not lift a finger to silence the *National Journal*, and Jackson never forgave him for that. It wiped away all his former respect for the President.[6]

Democrats countered by accusing Adams of pimping for the Czar of Russia when he was the U. S. minister to that country. In addition, they charged him with recklessly spending public money for gambling furniture in the White House, in particular a pool table, cues and balls. His lordly and aristocratic manner was projected in the press to emboss their image of a President who was hostile to the aspirations of the majority of the American people. But the most effective propaganda against Adams was the charge that he had conspired with Clay to steal the presidency from Jackson in 1825. The "corrupt bargain" cry echoed across the country. Its vibrations in Congress alone were loud enough to produce a bloodless duel between Henry Clay and John Randolph of Roanoke in 1826.

Jackson, on the other hand, was charged with adultery, seduction, murder, theft, treason, and other less strenuous crimes such as Sabbath-breaking, cockfighting, horse racing, and swearing. His duels with the Bentons and Dickinson were recounted with great relish and much distortion; his executions of militiamen were graphically described. One administration newspaperman by the name of John Binns hit upon an excellent idea of publicizing Jackson's butchery by printing a handbill, in which six black coffins were depicted representing six militiamen executed during the Indian wars. The Coffin handbill, as it was called, provided formidable propaganda for the administration forces in projecting the caricature of Jackson as a pitiless ruffian and murderer.

As this verbal brawl intensified, spraying the candidates with the stench of the gutter, two significant developments took place that would influence both major parties for the next several years. The first was the rise of Anti-Masonry in New York that began in 1826 with the disappearance and apparent murder of William Morgan, a stonemason from Batavia, New York. Morgan belonged to a local Masonic lodge and, because of a dispute with his lodge brothers, wrote a book disclosing Masonic secrets. When efforts to dissuade him from this betrayal failed, Morgan was arrested for indebtedness. His bail was paid, but as he left the jail he was seized and probably taken to the Niagara River, where he was drowned. His disappearance immediately engulfed the western districts of New York in wild excitement. Masons were excoriated for their membership in an organization committed to kidnapping and murder to advance its cause. Soon, the excitement was intensified by religious and psychological forces, by economic grievances, by abolitionism and temperance. The frustrated, the angry, the malcontent convinced themselves that Masons were intrenched in business, politics, and the courts—indeed, every important position within the power structure of the state—and were preventing non-Masons from getting ahead. Perceptive politicians such as Thurlow Weed of New York and Thaddeus Stevens of Pennsylvania quickly recognized the value of the uproar and began to direct it into the electoral field. At the height of the disturbance, someone discovered that Andrew Jackson was a "grand king" of the Masonic Order, and it almost finished him as a candidate in New York. Not even the discovery of Henry Clay's membership in the order deflected the animosity shown the Democratic party in those areas where Anti-Masonry spread.

The second significant development during the 1828 campaign was the rise of the Workingmen's Party in Philadelphia. This started when a group of laborers, intellectuals, physicians, merchants, speculators, lawyers, and others, organized to protest such abuses as the lottery system, the excessive distillation of liquor, and legislative aid to the monopolizers of the "wealth creating powers of modern mechanism." On the positive side, the Workingmen's Party advocated reforms in education, the auction system, imprisonment for debt, mechanics lien laws, monopoly, and paper money. Although it drew considerable strength from middle class capitalists, it also included a genuine labor group contending for the interests of the wage earner. It started in Pennsylvania in July, 1828, and spread into New York, Boston, Newark, Baltimore, and other eastern cities.[7]

As their organization improved over the final months of the campaign, the Democrats sought to direct the Workingmen's movement to their own purposes. They were not always successful in convincing reformers that their interest was altruistic; no matter, they more than made up for occasional failures by snaring large blocs of minority and majority groups through their highly organized party apparatus and the new, exciting techniques of electioneering that they introduced in the campaign. For example, they raised to a political art such things as rallies, parades, barbecues, and dinners. They began distributing buttons and special hats with hickory leaves stuck in them to add to the electioneering fun. In addition, the General himself was persuaded to hit the campaign trail, though he wisely disguised it to avoid shocking public decorum, which at that time regarded campaigning by a candidate as most improper. The occasion was the anniversary of the victory of New Orleans, and on the pretext of participating in a patriotic celebration, Jackson accepted an invitation to attend it. But as anyone with half an eye could see the event prodded the nation to remember (as though it could ever forget) the General's tremendous victory during the War of 1812, and it brought him the kind of attention and publicity that politicians soon learned could shape phenomenal majorities at the polls.

While Old Hickory paraded through New Orleans campaigning for the presidency, his friends in Washington were winning the election for him through legislation. Led by Van Buren, the Democrats in Congress concocted a tariff bill aimed at attracting electoral votes in both the Northeast and the Northwest by hiking the protective rates on items favored in those areas, such as raw wool, flax, molasses, distilled spirits, and hemp. After a bruising fight, in which subterfuge and deceit figured prominently, the tariff passed, and President Adams signed it. Although it was called the Tariff of Abominations in some quarters, especially in the South, the bill helped Jackson find votes in those states where he needed them most: Kentucky, Illinois, Indiana, Missouri, Pennsylvania, Ohio, and New York.

And find them he did. By the thousands, people crowded to the polls, guided by the Democratic Party. When the returns were tabulated in the late fall of 1828, Jackson had won 647,276 popular votes to Adams' 508,064. The number of voters who participated in this election nearly quadrupled the 1824 figure, and of these Jackson took approximately 56 per cent—a smashing victory that remained unequaled in the nineteenth century. In the electoral college the Hero received a majority of 178 votes, and Adams got 83. Every state south of the Potomac and west of the Alleghenies—including Pennsylvania—went for Jackson. Adams carried New England (with the exception of a single vote in Maine), Delaware, New Jersey, and most of Maryland. Together, the two candidates shared New York, with the General taking the larger portion.

Because he was the first western President of the United States, the very embodiment of the frontier, many people interpreted Jackson's election as the termination of government control by the planter and commercial aristocracy of Virginia and New England. Although the General himself was a Tennessee aristocrat of sorts his coming to power was expected to create new links between the central government and the large masses of citizens from every section of the country.

At his home in Nashville Jackson started receiving congratulations on his election early in the autumn when the first returns began to indicate a stupendous victory in the making. During the campaign, he had kept a close watch on many of the details that went into the construction of the Democratic party, supervising, directing, and suggesting what he thought would build a majority. Then, at the moment of his triumph, he sustained a crushing personal loss. His wife, Rachel, died suddenly and very tragically. She had a history of heart seizures. These were slight, but they intensified during the presidential campaign as the attacks on her character became more pointed and more cruel. On Wednesday morning, December 17, as she went about her household duties, she felt a knifelike pain in her chest. She clutched at her heart, "uttered a horrible shriek," and collapsed in a chair. For the next several days her husband watched at her bedside; on the twenty-third of December, because she rallied slightly, she was moved for the first time from her bed to a nearby chair in order to rearrange and smooth the bedsheets to make her more comfortable. While sitting in the chair, she was seized with a second attack. She let out a long, loud cry. Her head fell forward, and she "never spoke or breathed again." As she was carried back to her bed there was a wild rushing about of friends, relatives, and servants. Jackson, almost beside himself with fear, ordered the doctor to bleed her. The doctor opened her arm, but no blood flowed. In a panic, Jackson cried out, "Try the temple, Doctor." Again, the doctor operated. Two drops of blood oozed out and stained her cap. Nothing more could be done, but for a long time the General refused to believe she was really dead. All night long he sat by her side, his face in his hands. Not until Rachel's hands and feet grew cold was Jackson convinced that she was gone. The next morning, he was found by William B. Lewis in a state of absolute desolation. He was speechless and utterly inconsolable.

Rachel was buried on Christmas Eve in the garden of the Hermitage, after which Jackson wrapped himself in his grief and refused to be budged, either to leave his house and start for the capital or turn to the affairs of state that demanded his attention. It was weeks before he stirred about the house in any recognizable way. Then, slowly, his mind reached for the problems confronting his new administration, and he began to make preparations to leave for Washington. About the middle of January, on a Sunday, he boarded a ship at Nashville to start the trip eastward. He was accompanied by his nephew, Andrew Jackson Donelson, who would serve as his private secretary; by Donelson's wife, Emily; by William Lewis; and several others. As he boarded the ship, partisans placed hickory brooms on the bow and stern of the vessel, but the General paid them no heed. As the ship sailed up the river, other vessels, crowded with people, circled close by. The men called to Jackson with loud and sustained "huzzahs," and, although the President's heart was "nearly broke," something inside him responded. He left his cabin and came out on deck, saluting the people and tipping his hat to them. They replied with even louder cheers, and several remarked how wonderful it was to have at last "a man of the people" living in the White House.[8]

CHAPTER VI

Masterful Politics

THE INAUGURATION OF ANDREW JACKSON WILL ALWAYS REMAIN part of the folk legend of American history. People from every point of the compass converged on the capital to witness the final "triumph of the great principle of self government"—or so the Democrats contended. By the thousands they came, pushing into every avenue and street leading to the Capitol and reminding one observer of the inundation of the German barbarians marking the downfall of the Roman Empire. "I never saw such a crowd here before," exclaimed Daniel Webster. "Persons have come five hundred miles to see General Jackson, *and they really seem to think that the country is rescued from some dreadful danger!*" Indeed, they did: from the tripleheaded danger of aristocracy, privilege, and corruption. Like a mighty, agitated sea, the mobs hustled and shoved their way to a place in front of the spot where Jackson would appear and where the outdoor ceremony would commence. Eager to get closer, they swarmed over the steps leading to the portico so that a ship's cable had to be stretched about two-thirds of the way up the flight of stairs to contain them. Francis Scott Key, standing to one side near a gate of a large enclosed yard, gasped in wonder at the incredible spectacle of this surging, pulsating mass of humanity. "It is beautiful," he said, "it is sublime!"

Suddenly, there was a stupendous roar from the people as they spotted the Hero approaching between the columns of the portico. Men tore off their caps, and the color of the scene changed "as if by a miracle." The dark tints of a motley crowd on a late winter afternoon vanished, replaced by

[118]

the "bright hue of ten thousand upturned and exultant human faces, radiant with sudden joy." The shouting seemed "to shake the very ground." Numbering fifteen thousand, the mob chanted, waved, and saluted their champion. Jackson, deeply moved, bowed low before "the majesty of the people." At length the Chief Justice, John Marshall, stepped forward to begin the ceremonies, and after the oath of office was administered, the President kissed the Bible, which again activated the crowd. Then, when Jackson started to read his inaugural address, the audience tried to quiet down to listen but he spoke so softly that only those within a few feet of him were able to hear his words. For the most part, they just waited out the ten minutes or so it took him to finish his address.

What the President said was a model of political adroitness. It was short, vague, and contained nothing to frighten his friends. He straddled the questions of the tariff and internal improvements and said he would protect the rights of the states, "taking care not to confound the powers they have reserved to themselves with those they have granted to the Confederacy." One thing he sincerely wanted to do was extinguish the national debt, a liability "incompatible with real independence," he said. Also, he believed the government needed an overhauling, insofar as the patronage was concerned, in order to remove it from "unfaithful or incompetent hands." Every expectant officeholder in the crowd who could hear that remark must have quivered with joy and anticipation.

When the speech ended, a thunder of cannons sounded close by, echoed moments later by other guns at the Navy Yard and the Arsenal. The crowd closed in on the President in the hope of shaking his hand, and it was only with the greatest difficulty that he pushed his way through the mob to a carriage that was waiting to take him to the White House. Once aboard, a group of Revolutionary War veterans stationed themselves alongside the carriage as a guard of honor, and together they slowly moved down Pennsylvania Avenue, while Jackson, standing at his place, gestured his thanks to the people lining the route and applauding him as he rode by.

At the White House, preparations had been made for the President to meet the people informally. But what took place was never anticipated. An immense throng appeared of "all sorts of people, from the highest and most polished, down to the most vulgar and gross in the nation. I never saw such a mixture," said one man. "The reign of King 'Mob' seemed triumphant." Jackson shook hands with the well-wishers, but the press of the crowd threatened to injure him, and several gentlemen were obliged to surround him and shield him from the pushing and shoving. The pressure got so bad that the President finally ducked out a side entrance and fled to his lodgings at Gadsby's Hotel. The crowd did not pursue him but turned their attention instead to the food and drink, scrambling to grab the refreshments from the hands of the waiters, breaking china and glassware in the process, spilling pails of liquor on the carpets, and smearing mud all over the damask covered chairs as they climbed over them to get a better look. The destruction reached such a pitch that tubs of wine and ice cream were carried outside to the garden to draw the crowd out of the house, whereupon the guests dove through the windows to reach the prize. It was a "regular Saturnalia," laughed one Congressman who witnessed the mayhem. The mob, he wrote, "broke in, in thousands, Spirits black, yellow, and grey, poured in in one uninterrupted stream of mud and filth . . . many subjects for the penitentiary. . . ."[1]

In the eyes of a few this was a frightful beginning for a new administration, a portent of certain disaster. But for others, it marked the beginning of truly democratic administration, the first in American history, one blessed and baptized by the joyful screams of a wildly enthusiastic public.

The celebration over, Jackson returned to the partially wrecked White House and prepared to assume control of the government. To help him in his task he had chosen a Cabinet representing both the several factions of his party and the different sections of the country. In the first seat, as Secretary of State, he selected an estimable man, Martin Van Buren—dubbed the Little Magician or Red Fox of Kinderhook in tribute to his finesse in handling men and affairs. For balance—to satisfy the Calhoun wing of the party—Jackson gave the Treasury Department, the second most important position in the Cabinet, to Samuel D. Ingham of Pennsylvania. Then because he wanted someone in his official family who was a personal friend and upon whom he could rely for personal loyalty, Jackson invited John Eaton to take the War Department. For the Navy, he chose John Branch, a former governor of North Carolina, an intimate friend of Eaton, and a man whose cultivated manners would presumably add social distinction to the new administration. John McPherson Berrien of Georgia, a fussy and opinionated dandy, became the Attorney-General, and for what compelling reason he was selected had politicians scratching their heads. Again Eaton's influence was decisive, and about the best that could be said for Berrien was that he advocated the removal of the Indians west of the Mississippi. As for Postmaster General, the incumbent John McLean of Ohio, another of Calhoun's friends, was continued in the post he held under John Quincy Adams. But a short time later a vacancy occurred on the Supreme Court, and McLean asked to have the seat under the assumption that it was a safer and more comfortable place from which to launch his presidential ambitions. William T. Barry, the recently defeated candidate for governor of Kentucky, had been tagged for the slot, but it was rearranged so that McLean went on the Court and Barry took the Post Office Department instead.

Like most strong presidents, Jackson chose a Cabinet that was essentially weak, probably because he meant to dominate it. Only Van Buren was an officer of the first rank, the rest slid down the scale from average to barely passable. Nevertheless, Jackson was pleased with his balancing act and pronounced the Cabinet "one of the strongest . . . that ever have been in the United States. . . ." Of greater consequence in advising the President, however, was a group of men, called the Kitchen Cabinet, who had been most instrumental in arranging his election. Several of them, like Eaton, Lewis, and Donelson, were Tennesseans who had known Jackson and worked with him for many years; others were newspaper men from different states, among them Amos Kendall, the wizened, prematurely gray editor of the Kentucky *Argus of Western America,* who had labored furiously and successfully in Kentucky to unite the Relief and Anti-Relief factions of the party behind the General. As reward, he was appointed Fourth Auditor of the Treasury. Another member of this Kitchen Cabinet was Isaac Hill of New Hampshire, editor of the *Patriot,* a small, lame, cadaverous looking man, who had built a political machine to "revolutionize" the politics of his state for Jackson. He was named Second Comptroller of the Treasury, but when the Senate refused to confirm him he achieved a kind of revenge by winning election to the Senate itself. A third member of the Kitchen Cabinet was Francis P. Blair, a mousy looking man, hardly weighing more than a hundred pounds who was summoned to Washington from Kentucky, where he had been Kendall's associate, to edit the Washington *Globe.* He founded this newspaper on December 7, 1830, to replace Duff Green's *Telegraph* as the official party organ. Blair (the President pronounced it "Bla-ar") was an exceedingly provocative editor whose articles were usually laced with political invective. Naturally Jackson had profound respect for such a skillful master of polemics and accorded him a place close to his ear. Other members of the group included Roger B. Taney, a leading Democratic organizer in Maryland and later Attorney-General, Secretary of the Treasury, and Chief Justice of the United States; and also Martin Van Buren, in many ways the most influential of all, once the President measured his loyalty and learned to appreciate his singular talents. Not that Jackson ever relied absolutely on any one man or group of men. Indeed, the term Kitchen Cabinet connotes greater importance than it had. The President sought advice from these men, listened to what was proposed, but invariably made up his own mind, not a few times contradicting what his closest advisers recommended. To regard the Kitchen Cabinet as the power behind the throne, dictating policy or establishing guidelines for political action is a gross error. Rather, it was an informal, changing corps of men, functioning under the direct supervision of the President, whose ideas and opinions, verbal and written, aided Jackson in his search for a more aggressively democratic program of reform.

Even in the matter of patronage, where the members of the Kitchen might hope to exercise control, their influence was feeble. Most of them expected and desired wholesale removals, only to discover the President committed more to the *principle* of rotation than to its actual practice. And there could be no doubt of Jackson's determination to establish the principle. No one had a perpetual right to office, he declared, and those who thought otherwise would be among the first to leave government service. The opinion of the men in the Kitchen was probably best expressed in a Senate speech delivered by William L. Marcy, one of Van Buren's henchmen. "To the victor belong the spoils of the enemy," he said; yet despite this lovely promise Jackson removed only 919 persons out of 10,093 during the first eighteen months of his administration. For the entire eight years of his presidency he removed only nine per cent of all officeholders, which is approximately one in ten. Certainly this was not the record of a spoilsman, particularly when the figures are considered in the light of normal replacements due to death and resignation, plus those dismissed for incompetence and dishonesty. Still Jackson was determined to establish an important democratic principle of government. Unfortunately, rotation gave him a bad name, for he did not dissociate himself from those like Marcy who talked of spoils as the proper rewards of election. Also, he made a few disastrous appointments like Samuel Swartwout, the fast-stepping embezzler of the New York Customs House. Nevertheless the principle of rotation was sensible and desirable in this period of rapid organization of parties. For one thing, it advanced the democratization of government by easing the method by which new men, sympathetic to the views of an incoming administration, could be readily assigned to offices held by appointees of a previous administration. The system has obvious disadvantages when applied by incompetents, political hacks, and crooks, but, properly controlled, it provides fast response to the changing demands of the American people as expressed through the ballot box.[2]

Troublesome to Jackson though it was, however, the spoils system did not produce nearly the scandal that boiled out of the Eaton Affair—or "Eaton Malaria," as Van Buren called it. This wretched business developed because John Eaton, a widower, had married the daughter of a Washington innkeeper, the pretty, witty, saucy—and sometimes nasty, if her slave is to be believed—Margaret (Peggy) O'Neale, who frequently served as a barmaid in her father's hotel. Peggy herself was a widow who had married John Timberlake, a purser of the U. S. Navy. While at sea, Timberlake cut his throat in a fit of melancholy brought on by drunkenness and the knowledge that he had been cuckolded by Eaton. When the dreadful news of Timberlake's death reached Washington, Eaton had some kind of fit of his own and decided to marry Peggy. But before doing so, he asked Jackson for an opinion. "Why, yes, Major," replied the Hero in his best marriage-counselling voice, "if you love the woman, and she will have you, marry her by all means." A little nervously, Eaton responded that Peggy's reputation in Washington had not escaped reproach. "Well," Jackson fired back, "your marrying her will disprove these charges, and restore Peg's good name." And so, on January 1, 1829—two months prior to the inauguration—the restoration took place. But the wedding did not disprove the charges; in fact, in the minds of some, it proved them, and Congressmen started alluding to it with some pretty crude remarks. "There is a vulgar saying of some vulgar man, I believe Swift," wrote the amused Churchill C. Cambreleng, Representative from New York, "on such unions—about using a certain household . . . and then putting it on one's head." Because Eaton was a member of the Cabinet, walking about in the presence of polite society with this ". . . " on his head could prove very embarrassing. In no time at all Floride Calhoun, wife of the Vice-President, said it was disgusting, and that she had no intention of associating with the likes of Peggy Eaton. But this was strange indeed, for Mrs. Calhoun visited Peggy immediately after the marriage! Why this sudden change of mind? Obviously it had nothing to do with the marriage. What brought about the change was the announcement of the Cabinet appointments. Floride's about face simply reflected her husband's resentment and jealousy of Eaton's influence in the selection of Jackson's official family. This factor is pivotal in assessing responsibility for the sordid events that followed. For the Vice-President later tried to duck away from the consequences of his folly by blaming his predicament on his wife's adamant refusal to meet the Eatons as social equals.

Another lady who took exception to the nuptials was

Emily Donelson, the wife of Jackson's nephew and secretary. But, again, her reaction was the result of her husband's admitted indignation over Eaton's influence with his uncle. (Snorted Donelson, "It is a pity that Eaton was brought into the Cabinet.") Both ladies, in serving the ambitions of their husbands, sought Eaton's disgrace through the vulnerable and pathetic little figure of his wife.

Thus, at the commencement of his administration, Jackson found himself enmeshed in a conspiracy to injure "a helpless and virtuous female." At official state functions he noticed with alarm how intense and concentrated was the feeling against Peggy. Naturally his wrath blazed up against those malicious gossips, for he saw in her something of the dead Rachel, whose sainted life, he said, was shortened by just such slander and calumny. But Jackson was no fool. He probably had a suspicion of what might happen when he invited Eaton into his Cabinet, since he was no stranger to this sort of attack; still he fully expected responsible officials to act like gentlemen and accept his decision once the appointment was conferred—or, as he told John Randolph, to remove themselves from the Cabinet if they could not serve with Eaton. Instead, things went from bad to worse. They finally slipped out of hand and provoked the President into a passionate defense of Peggy when the Reverend J. N. Campbell, pastor of the Presbyterian Church in Washington where Jackson and Rachel had been parishioners, contacted the Reverend Ezra Stiles Ely of Pennsylvania and urged him, when he visited Washington for the inauguration, to speak to Jackson about the unfortunate appointment. Ely was an old friend and presumably could talk to the General about so delicate a subject without setting off an explosion. Unfortunately Ely missed his chance in the capital, and therefore, when he returned to Philadelphia, he wrote Jackson a long letter reciting the gossip he had heard about Peggy: that she was "dissolute," that she instructed her servants to call her children Eaton and not Timberlake, that Rachel herself regarded Peggy as a fallen woman, and so on. Jackson's reply was typical, direct, and deeply felt. He dismissed the charges against the woman as revolting slander, initiated by that prince of slanderers, Henry Clay—or so a "very intelligent lady" informed him. This charge against Clay was vicious and false but one to which Jackson was understandably susceptible. As for Rachel's supposed condemnation of Peggy, that was simply not true. "Mrs. Jackson, to the last moment of her life, believed Mrs. Eaton to be an innocent and much injured woman," wrote the President. Furthermore—and this point Jackson would reiterate to several friends—both Eaton and Timberlake were Masons, and neither could "have criminal intercourse with another mason's wife, without being one of the most abandoned of men." In a more poetic mood, for Jackson was a great romantic, he said, "Female virtue is like a tender and delicate flower; let but the breath of suspicion rest upon it, and it withers and perhaps perishes forever. When it shall be assailed by envy and malice, the good and the pious will maintain its purity and innocence, until guilt is made manifest—not by *rumors* and *suspicions,* but by facts and proofs brought forth and sustained by respectable and fearless witnesses in the face of day."[3]

In order to hear "facts and proofs" by respectable witnesses, Jackson called a Cabinet meeting on September 10, 1829, to which Dr. Ely and the Reverend Mr. Campbell were invited to attend to present their evidence. Eaton was noticeably absent. Jackson opened the meeting with a statement on the meanness of calumny against which Christian men and women must constantly war. Next, he denied a charge leveled by the Reverend Mr. Campbell that Peggy had had a miscarriage during one of Timberlake's extended leaves from home. Then he turned to Dr. Ely and asked him whether he had turned up relevant information during his investigation of the affair. Ely admitted there was no evidence to convict Eaton of improper conduct. "Nor Mrs. Eaton, either," snapped Jackson. "On that point," replied Ely sternly, "I would rather not give an opinion." "She is as chaste as a virgin!" roared the President. Ely blinked in disbelief but said nothing, whereupon Campbell was asked to make a statement. Speaking rapidly, he insisted that all he had tried to do was spare Jackson's administration from reproach and the "morals of the country from contamination." Jackson listened to this talk for as long as he could stand it and finally cut him short with a command to *give evidence*," not make a speech. Insulted, Campbell gathered up his papers and walked out. And with that, the President adjourned the meeting.

It was serious enough that two supposedly respectable clergymen were intent on hunting down a government official in the name of virtue and good morals, but it was intolerable when the wives of several Cabinet members joined the hunt by publicly insulting Peggy and refusing to associate with her. The purity brigade was led by Mrs. Calhoun and Mrs. Ingham, whose social standing was so superior to everyone else's that they had no difficulty in intimidating and then enlisting the aid of Mrs. Berrien and Mrs. Branch, both of whose husbands owed their Cabinet places to Eaton. In addition, several wives of the foreign envoys, particularly Mrs. Huygens, the wife of the Dutch minister, declined to receive Mrs. Eaton. But the final blow was the refusal of Emily Donelson, mistress of the White House, to call on Peggy as her duty required. She would receive the lady in the Executive Mansion as her uncle's guest, she said, but she would not return the courtesy. Deeply offended by her action, Jackson ordered her back to Tennessee. "I was willing to yield to my family every thing but the government of my House and the abandonment of my friend without cause," wrote Jackson. Donelson returned home with his wife, but six months later mutual friends effected a reconciliation between the President and his family, and the Donelsons returned to Washington in September, 1831.

This extraordinary affair, so ridiculous on the surface, actually masked a deeper political conspiracy instigated by the Vice-President, John C. Calhoun. It originated with Calhoun's growing jealousy of Martin Van Buren, and his need to challenge his rival's sudden ascent. Both were ambitious for the presidency and wanted it after Jackson, and neither had the patience or grace to wait until the other had his turn. Thus, when the Cabinet was being filled both men expected to dominate it; both were grievously disappointed by the final result, Calhoun the more so because there was "no one from South Carolina" in the Cabinet, and because his friends made a special effort to keep Eaton, his critic, out of it and have either Robert Y. Hayne or James Hamilton, Jr., appointed in his place. The failure, therefore, was two-fold: not only was Eaton in, but his influence was clearly visible in the appointment of several others. With his position and possibly his future thus jeopardized, Calhoun recognized he must drive Eaton from the Cabinet or see his own presidential ambitions sustain a decisive setback.

The method of assault proved easy. Eaton had just married the notorious Peggy. Although both Mr. and Mrs. Calhoun had visited the newlyweds before the Cabinet appointments were announced, they pretended it never happened in the general uproar they instigated over the necessity of accepting Mrs. Eaton as a social equal. Calhoun figured that if the ostracism continued long enough, Eaton in time would buckle under the pressure, resign his post, and be replaced by Hayne, Hamilton, or some other South Carolinian. Yet the danger of reminding Jackson of Rachel's agony was not lost on Calhoun, and like his previous hypocrisy in the Monroe Cabinet he disguised his true intentions, pretending he was powerless before his wife's unbending interdict of Peggy. But this was obviously false, since the good lady had already paid her respects to Mrs. Eaton at least once before, shortly after the wedding.

Actually, the Secretary of War was not really as dangerous as the Vice-President pretended, since Eaton was extremely vulnerable. The real danger, as Calhoun knew, was Martin Van Buren—and Van Buren was neither vulnerable nor stupid. He was a widower, with no daughters to embarrass him in the social turmoil the Calhouns had created. He had only sons, four of them. Thus, because he was a clever politician, alert to any advantage which opened up to him, he repeatedly called on Mrs. Eaton and showed her the respect she was entitled to as the wife of a Cabinet officer. In addition, to exploit his advantage even further, he persuaded Sir Charles Vaughan, the British minister, and Baron Krudener, the Russian minister, both bachelors, to treat Peggy with courtesy and attention. At several dinners, one given by Van Buren and another by Krudener, these gentlemen performed manfully, but the wives of the Cabinet members refused to attend the affairs and so the simple politeness shown on the one hand was matched with cruel snubs on the other. At Krudener's dinner, the wife of the Dutch minister put in an appearance, but when she found she was assigned to a seat next to Mrs. Eaton, she was so offended that she grabbed hold of her husband's arm and stalked out. In her anger, she even threatened to give a ball from which Peggy would be purposely excluded, and the Mesdames Branch, Berrien, and Ingham chorused a similar intention.

Jackson, infuriated by this affront to the Eatons, seriously considered ordering the Dutch minister home. "What divine right," he stormed, "let females with a clergyman at their head have to establish a secrete inquisition and decree who shall, & who shall not, come into society—and who shall be sacraficed by their secrete slanders. . . ." But at least there was one gratifying aspect to the unseemly business, and that was Van Buren's correct and gentlemanly behavior toward Peggy, which Jackson appreciated. Others seemed intent on embarrassing his administration; not so the Secretary of State. And, of course, the more Van Buren treated Peggy courteously the more Calhoun schemed to humiliate her. "Calhoun is forming a party against Van Buren," wrote Daniel Webster, "and as the President is supposed to be Van Buren's man, the Vice President has great difficulty to separate his opposition to Van Buren from opposition to the President."

The mistake that Webster made was also Calhoun's mistake: thinking Jackson was Van Buren's man, that Van Buren in any way controlled or determined the policy of the administration. Jackson had profound regard for Van Buren's capabilities, but he was his own man, something the Red Fox discovered within a month after his arrival in Washington. Unfortunately for the Vice-President, he never learned this, and the disasters he called down on his own head he later excused by attributing them to the despicable machinations of the Little Magician.[4]

That Jackson was his own man can be seen by his steady pursuit of his own democratic aims, a pursuit that became more aggressive as his administration advanced into the decade of the thirties. His first message to Congress, for example, hinted broadly at the direction he proposed to follow, both in foreign and domestic affairs. In foreign affairs, he assumed the posture of a fierce nationalist, stating his determination to "cause all our just rights to be respected" and, where differences existed, to speed them to a conclusion favorable to the interests of the United States. Muscular patriots always found Jackson a President to delight their souls, for his words and actions bristled with defiance. But more important than his theatrics was the way in which he exercised his executive powers to initiate a forceful foreign policy. Under Jackson, the presidential office became a more active and aggressive agent in the resolution of diplomatic problems.

In domestic affairs, his language was less belligerent. Advocating democratic reform of the electoral system, he proposed a constitutional amendment to guarantee that the presidency be reserved only for those persons who had won the "fair expression of the will of the majority," thus preventing a recurrence of the 1824 "corrupt bargain." He also suggested limiting the presidential term to four or six years. Then, on rotation of office, he asserted the principle as forcefully as he knew how, presenting it as another expression of the nation's advancing democracy. "In a country where offices are created solely for the benefit of the people," he wrote, "no one man has any more intrinsic right to official station than another. Offices were not established to give support to particular men at the public expense. No individual wrong is, therefore, done by removal, since neither appointment to nor continuance in office is a matter of right."

Continuing his message, Jackson touched on a matter that he repeatedly cited as one of the first goals of his administration: the extinguishment of the national debt. To succeed in this aim, he said, would galvanize the economy with "additional means for the display of individual enterprise." If that were done, he thought it possible to distribute the surplus revenues of the federal government to the individual states.

Next, the President brought up the Indian problem. Literally sighing with each word (their present condition "makes a most powerful appeal to our sympathies," he said) Jackson contended that it had long been the desire of the government to introduce the arts of civilization among the tribes and to turn them from their wandering life. They had been pushed "from river to river and from mountain to mountain, until some of the tribes have become extinct. . . ." As one who had done a great deal of pushing to advance the Indians along their road to extinction, Jackson now proposed to do something to reverse the course by suggesting to Congress "the propriety of setting apart an ample district west of the Mississippi . . . to be guaranteed to the Indian tribes as long as they shall occupy it. . . ." This emigration, he averred, would be voluntary, "for it would be as cruel as unjust to compel the aborigines to abandon the graves of their fathers and seek a home in a distant land."

Turning to the tariff, the President handled it gingerly because of its controversial character. Southern reaction to the Tariff of 1828, including public threats of disunion, proved how dangerous it was. The issue fissured the nation along sectional lines, with the Northeast and Northwest advocating protection for manufactures and certain raw materials, and the South damning it as a government subsidy paid to Northerners out of the pockets of Southern planters and farmers. Like the excellent politician he was, Jackson advocated compromise (the "middle course" he called it) in order to "avoid serious injury and to harmonize the conflicting interests of our agriculture, our commerce, and our manufactures." He invited the Congress to examine the present law with the object of modifying "some of its provisions. . . ."

Then, in two small paragraphs—seventeen lines, to be exact—he took notice of the mammoth Second Bank of the United States, that financial colossus whose power to initiate financial catastrophe had already been demonstrated in the Panic of 1819. He reminded the Congress that the Bank's charter would expire in 1836, and that undoubtedly its stockholders would apply for "a renewal of their privileges." In justice to all, he said, an early consideration of the application might be appropriate, inasmuch as "a large portion of our fellow-citizens" question both the constitutionality and the expediency of the present charter; "and it must be admitted by all," he continued, "that [the Bank] had failed in the great end of establishing a uniform and sound currency." Thus did Jackson hint at a possible war with the Bank, placing it on notice that he did not like the way things were presently constituted, and that he wanted changes.

On the face of it, this presidential program seemed modest enough. In essence, Jackson called for strengthening the national posture in foreign affairs, eliminating the national debt and releasing surpluses to re-energize business, further democratizing of the government through rotation, compromising the tariff to restore harmony between the sections, moving the Indians west of the Mississippi River, and altering the existing charter of the Second Bank of the United States to eliminate constitutional objections that had existed since the Bank's inception. It is rather remarkable how much of this program Jackson actually accomplished. Unlike previous presidents, he did not shape a program and then wait for the Congress to act upon it. Left to its own devices Congress frequently emasculates a presidential program. Instead, Jackson initiated very close working relationships with both houses of the legislature, and in this he had the assistance of a party he had built along with top flight politicians like Benton, James K. Polk, Van Buren, Roger Taney and others who understood the need for cooperation between the executive and legislative branches of the government. But it was Jackson himself who suspected and explored the true possibilities of his office. Earlier presidents were hobbled by their concern for the separation of powers; not so Jackson. He purposely set out to bridge the separation in order to persuade the Congress that what he wanted by way of legislation actually represented popular will. He forged an image of a strong executive, acting for the benefit of all the people: planter, farmer, merchant, mechanic, laborer, and manufacturer.[5]

Of all the issues touched on by Jackson in his first message to Congress, the tariff was the one most likely to spark a controversy. It aggravated sectional difference, so much so that after the passage of the 1828 tariff, John C. Calhoun wrote an *Exposition and Protest,* which argued the constitutional right of the state to nullify the tariff if not repealed. South Carolina withheld applying the Calhoun doctrine because it presumed that Jackson, as President, would redress its grievances and repeal the law. But events did not work out that way, and a stormy debate erupted in the Senate between Robert Y. Hayne of South Carolina and Daniel Webster of Massachusetts which warned of an increasing divergence of opinion over the right of the state to protect itself and over the nature of the Union. The debate began not with a tariff discussion, however, but when Senator Samuel A. Foote of Connecticut moved to limit the sale of public lands—a proposal that promptly brought a howl from westerners who were sensitive about migration. Senator Benton of Missouri bolted to his feet to challenge the proposal, and he ended his remarks by warning the east against further harassment. Daniel Webster replied to this outburst, after which Robert Hayne rose to defend his western allies. The ensuing debate struck off conflicting interpretations about cohesion within the Union, with Webster insisting that union and liberty were inseparable and Hayne, in a closely argued speech, defending States' rights and enunciating the Calhoun doctrine of nullification.

Although Jackson subscribed to a very fuzzy kind of States' rights philosophy (actually it was derivative to begin with and had been weakened considerably by the time he became President because he felt the states had nothing to fear while he was in the White House), he did not believe that the privileges of the states included threatening the existence of the Union. States had rights as long as they acknowledged their bond within an indestructible union. Let them threaten that bond in any shape or form, and he would teach them the meaning of treason. He was given an opportunity to erase doubts about his position on the question when the Democratic faithful held a commemorative celebration on April 13, 1830, to honor the birthday of Thomas Jefferson. Advocates of nullification planned to use the occasion to enlist the support of other states in their extreme interpretation of States' rights. And on that point Jackson meant to stop them. After receiving his invitation to the dinner he consulted with Van Buren, now his close adviser by virtue of the Magician's loyalty, intelligence, and circumspection. They agreed that Jackson's duty lay in informing the nullifiers as well as the other members of his party precisely where he stood; so the President went to the dinner invigorated with a sense of impending battle.

After dinner, Jackson was called upon to propose a toast. Rising in his place he fired his volley squarely into the faces of the unsuspecting "nullies." "OUR FEDERAL UNION," he said, "IT MUST BE PRESERVED." The blast ricocheted around the room. There was no question of its meaning or Jackson's mood. It had the punch of a political slogan and the impact of truth.

When his turn came to propose a toast Calhoun tried to recover something for the nullifiers by advocating, "The Union—next to our liberty the most dear; may we all remember that it can only be preserved by respecting the rights of the States and distributing equally the benefit and burden of the Union." But Calhoun's toast did not have the force and thrust of Jackson's seven-word manifesto. It went on too long. He should have stopped after "liberty most dear." That was another of Calhoun's faults; he never knew when to stop talking.[6]

Although Calhoun's habit of running off at the mouth along with his fumbling politics converged to shape his political demise, it is most probable that Jackson had already marked him for the political junkheap. It developed this way. Between 1823 and 1828 the General had been shown various bits of evidence by such confidants as William B. Lewis and Sam Houston documenting Calhoun's vote of censure during the time of the Seminole controversy. For this treachery, the conspirators insisted on the South Carolinian's ouster from the Jackson camp. Whether this evidence convinced the Hero is difficult to tell; probably it did; nevertheless he said nothing and did nothing because there was nothing he could do since his election might depend on the support he received from the Calhoun forces throughout the country (especially in the South), and he was too smart a politician to throw that support away. Nor, after the election, did he intend to cripple his administration at the very outset by brawling with his Vice-President over an incident that had occurred ten years before. His pride may have been bruised a little, but he would never sooth it with a pointless quarrel that could wreck his administration. Later, maybe, when it was all over, he would carry on in his usual melodramatic fashion, repeating endlessly how he had been betrayed and abused, how his loyalty had been repaid with deceit; but for the moment he was circumspect, and it is very possible that if Calhoun and his friends had left the thing alone Jackson would have gone on as if nothing had happened—just as he had always done. Instead, they continually provoked him. They not only insisted on asserting control over the Cabinet and the administration by plotting Eaton's removal, but they made their prosecution personal by dividing Jackson's own family and forcing him to order the Donelsons home to Tennessee. Still as these events unfolded, the President continued to exercise extraordinary restraint. Although he complained about his troubles in private, that was as far as he went. Even after the Jefferson dinner, when the President made it abundantly clear that the principles of nullification were an abomination to him, he did not initiate a break with the Vice-President. Nor did he purge him when he received a letter from William H. Crawford which, in Jackson's words, proved "Calhoun a *villain*." The letter was firsthand testimony that the South Carolinian voted for Jackson's censure in Monroe's cabinet; after reading it, the General passed it along to the Vice-President with a short note asking if the charge was correct. Responding in a fifty-two page letter, Calhoun tried to excuse his artless dodging as Secretary of War, but he left no doubt about his earlier hypocrisy. The President let it pass with a curt reply that "no further communication with you on the subject is necessary." For the remainder of the spring and summer of 1830 Jackson carried on a reasonably proper correspondence with the Vice-President, obviously trying to avoid any further discord. Had he been intent on an open rupture with Calhoun he had more than sufficient cause by the fall of 1830, but he wisely held his peace and did nothing to provoke a split.[7]

The break was brought about by Calhoun himself—and he did it by again letting his tongue run away from him. Now that he saw his succession to the presidency swiftly disappearing up the Hudson River toward Kinderhook, he abandoned all caution and launched a full-scale attack against the administration. He committed the colossal blunder of going to the people with his complaints. On February 17, 1831, he published a pamphlet in the United States *Telegraph* entitled, "Correspondence between General Jackson and John C. Calhoun . . . on the subject of . . . the deliberations of the cabinet of Mr. Monroe on the occurrences of the Seminole War." He not only prattled on and on about who said what about the censure, but he unwisely reviewed in public the scandal of the personal feuding that had been going on in the Cabinet for the past two years. In effect, he held up the Jackson administration to public ridicule. Naturally he claimed innocence of any wrongdoing; the villain of the piece, he shrilled, was that Red Fox of Kinderhook, that Little Magician, Martin Van Buren.

Publication of the pamphlet produced a violent reaction within the Democratic party and within the country at large. It was a ghastly error on Calhoun's part. No politician with any sense embarrasses his party in public this way, not if he wants support and not if he expects to continue in politics. Calhoun had committed the unpardonable sin of washing dirty linen in public, and since it had a ten-year history it was one of the less fragrant loads of wash ever carted out for the people to inspect. Jacksonian newspapers were appalled by the folly of the Vice-President; they accused him of wantonly disturbing Democratic harmony, of deliberately seeking Jackson's humiliation because of his own thwarted ambition.

Since Calhoun's action constituted a public attack, Jackson was outraged by the pamphlet. He composed a lengthy reply, but, on second thought, wisely allowed himself to be talked out of publishing it. Actually there was nothing further he needed to do. With great skill, Calhoun had performed a spectacular job of self-slaughter. The people and the party were all on Jackson's side, comforting him with expressions of loyalty and devotion.

Had the split come earlier, and under different circumstances, it might well have inflicted irreparable damage on the Democratic party. As it was, Jackson derived all the political advantages. By biding his time and acting without haste, he succeeded in getting rid of a man who was a threat to his administration, and whose ideas about nullification and States' rights were dangerous to the Union. Moreover, he did it without risking his popularity or splitting the party.

What about Van Buren? What was his importance? He desperately wanted to follow Jackson as President and undoubtedly manipulated everything possible to widen the rift between the two men. Furthermore, his very presence in the Cabinet goaded the Vice-President into a foolish test of strength. But the Magician's role in the affair was really quite incidental. The two protagonists were Jackson and Calhoun; the essential struggle was between the President and Vice-President to decide who controlled the administration and the succession. And, from the start, there was never any real doubt about how the controversy would end.

All that now remained was the physical removal of Calhoun's political corpse—and those of his friends in the Cabinet. What had to be averted, naturally, was a skirmish in Congress with the remnants of the Calhoun faction. But as long as Van Buren stayed in the Cabinet such a scrap could scarcely be avoided. So, bursting with loyalty, good sense, and an appreciation of what was expected of him, Van Buren offered himself for sacrifice. According to the Magician, he subtly worked out the arrangement for his final act of devotion during a riding trip with the President through Georgetown. Jackson had just expressed the hope that his administration might at last find peace, when Van Buren replied, "No! General, there is but one thing can give you peace." "What is that, Sir?" asked the President. "My resignation!" replied the Secretary. "Never, Sir!" the General responded solemnly, "even you know little of Andrew Jackson if you suppose him capable of consenting to such a humiliation of his friend by his enemies." For four hours, Van Buren explained why such a step was necessary and when the President finally came around to this point of view, he asked the Magician what he would do after his resignation was accepted. Return to law practice was the response; but Jackson vetoed that idea because it would constitute a triumph for both their enemies. Instead, it was agreed that Van Buren would be appointed minister to Great Britain. First, however, he would submit his resignation which would get things started; then Eaton would submit his, followed by the rest of the Cabinet. Once they were gone, the President could rebuild his Cabinet, eliminating completely the influence of John C. Calhoun.

Such is Van Buren's story about how the Cabinet was reshuffled, taking credit for suggesting his own resignation and convincing Jackson of its necessity. He may indeed have proposed his resignation, but it is hardly likely that he had to argue the President into something that was perfectly obvious. Van Buren had to go if the other members of the Cabinet were to be removed without provoking a fight in Congress. The only problem for Jackson was the avoidance of defeat through Van Buren's ouster, and that problem was adequately resolved by sending him to Great Britain.

In point of actual time Eaton's resignation preceded Van Buren's. The War Secretary insisted upon it. He submitted his letter on April 7, and Van Buren followed on April 11, 1831. With these in hand, Jackson demanded resignations from the rest. Said he: they came in together so they will exit together. Simple as that. Because Barry entered the Cabinet slightly later than the others and was a trusted member of the administration, he was permitted to retain the office of Postmaster General. For the rest, the reshuffling went off without serious incident. Van Buren went to England replacing Louis McLane, who returned to take over the Treasury Department; Eaton was replaced by Lewis Cass in the War Department; Edward Livingston became the new Secretary of State; Levi Woodbury of New Hampshire took the Navy Department; and Roger B. Taney was selected Attorney-General. Presumably, these changes would unite the Cabinet in loyalty and dedication to the President.[8]

It was masterful politics. The Calhoun faction, once so important to the Democratic cause, and now so dangerous, was stripped of its power. What is more, the stripping took place at a carefully selected time, just a month after the adjournment of Congress in 1831. The time chosen was long enough after the publication of Calhoun's pamphlet, commented one writer, for Democratic newspapers to complete their denunciation of its author, and for the issue to have subsided as an exciting topic of conversation and controversy. Furthermore, it came during the "lull" preceding the excitement of the next presidential contest. Since it also came nine months before the Senate would reconvene, confirmation of the new Secretaries would presumably pose no problem for the administration. Indeed, marveled one writer, "we may truly say of this disruption of the cabinet in 1831, that of all known political management it was the consummate stroke. Jacksonian boldness united with Van Buren tact could alone have achieved it."

But that was not the end of it. A few weeks later, Duff Green, editor of the *Telegraph,* stooped to a new political low by revealing in his newspaper the disgraceful stories about poor Peggy. He reprinted narratives of the scandal allegedly authored by Ingham, Branch, and Berrien. In fury, Eaton wrote Ingham asking whether he had sanctioned the article. Ingham retorted that the letter was "too absurd to merit an answer," whereupon Eaton demanded satisfaction. Ingham refused, pleading with Jackson to call off the madman and finally deciding that Washington was not a safe place to live with an outraged husband loose in the streets. He fled the city "in terror," reported the Washington *Globe,* thus marking the complete and utter rout of the Calhoun party.

In this new round of accusations, the Eatons were struck several fearful blows—Peggy most of all. And she was clever enough to recognize that the wrong did not rest with one person alone. She showed how perceptive she was when visited not long after by Jackson and Van Buren who had gone for a walk and suddenly found themselves before the lady's house. They decided to pay her a call, but she treated them with such formality and coldness—if not rudeness—that there was no mistaking her mood. What surprised Van

Buren particularly, he later recalled, was that "the larger share of the chilling ingredient in her manner and conversation fell to the General." Apparently, she felt she and her husband had been sluiced down the drain to serve the political needs of Andrew Jackson.

"There has been some mistake here," Van Buren whispered to the President when they rose to leave. Jackson merely shrugged as he went out the door.[9]

CHAPTER VII

"I Will Die with the Union"

WITH THE REMOVAL OF THE CALHOUN INFLUENCE IN GOVERN-
ment the Jackson administration began to move in a forward direction. A positive program of governmental action
slowly evolved, although some of its earlier aspects
appeared to critics as more negative than positive. During
the first years of his presidency, Jackson's own thinking
about issues and policies tended to be neo-Jeffersonian and
conservative, leaning toward States' rights and the econom-
ics of laissez faire, but fundamentally pragmatic in concept,
and suffused with a strong sense of popular need. Later in
his administration he edged closer to the notion of a strong
central government, but as he moved he was invariably
motivated by sheer political necessity. Consequently whenever there was disagreement between sections on practically
any political issue, Jackson tended to equivocate and, where
possible, to seek a compromise. For example, the question
of federally-sponsored internal improvements provoked
sharp differences of opinion. The National Republican party
of Clay, Webster, and Adams supported the principle in general. Since westerners also favored public works, the Hero
deliberately fuzzed his position on the issue, rather than
risk western hostility. Yet he was not equivocal when it came
to the Maysville Road, a stretch of proposed highway from
Maysville to Lexington within the state of Kentucky, the
home of Henry Clay. Friends of the projected road argued
that it would be an extension of the National Road and
therefore not something local in character; but Van Buren,
when still Secretary of State, denied this and characterized

the bill as purely intrastate and therefore unconstitutional. It is not surprising that the Magician argued this way, since his own state had constructed the Erie Canal at its own expense and did not relish the idea of the federal government aiding other states to build similar improvements.

Despite such objections, and the inflexible opposition of States' righters, the Maysville Road bill passed Congress in May, 1830. Jackson vetoed it. In his message, he not only negated the Maysville Road, but he challenged the principle that internal improvements were a federal responsibility. "If it be the wish of the people that the construction of roads and canals should be conducted by the Federal Government," he wrote, "it is not only highly expedient, but indispensably necessary, that a previous amendment to the Constitution, delegating the necessary power and defining and restricting its exercise with reference to the sovereignty of the States, should be made." What gave the veto special significance was the apparent hostility of the administration to the concept of an energetic central government in Washington—that it was reacting from a basic commitment to States' rights. But this was false. Jackson was not as much opposed to public works as his message implied, and he later approved several bills authorizing federal assistance for internal improvements. Fundamentally, the purpose of the veto was twofold: kill this particular bill because of its connection with Clay; and slap down what Van Buren called the "Internal Improvements party" (otherwise known as the National Republican party), which opposed his administration.

There was a tremendous uproar in the country over the veto, much of it organized by Clay and much of it ridiculous wailings about the dismal future of public works. But the message, recorded Van Buren, was "the entering wedge" by which "the Internal Improvement party was broken asunder and finally annihilated." It was the first significant lunge against "a triumvirate of active and young statesmen" who were seeking "power . . . to achieve for themselves the glittering prize of the Presidency, operating in conjunction with minor classes of politicians . . . and backed by a little army of cunning contractors. . . ." The veto, in other words, was another of Jackson's carefully calculated political moves— not an expression of administration philosophy to be taken as immutable party doctrine. Although he continued throughout his administration to fret over the constitutional difficulties involved in public works, Jackson never regarded the Maysville veto as an absolute canon against improvements—at least not when they were sponsored by loyal Democrats. Said he to Congress, the first time both houses were won by his party, "I am not hostile to internal improvements and wish to see them extended to every part of the country."[1]

Another instance of Jackson's apparent commitment to the States' rights position was his treatment of the Indians. Many of the problems respecting the Indians sprang from the white man's greed and racist disregard of Indian rights. The steady push of Americans west and south eventually cornered approximately fifty-three thousand Cherokees, Creeks, Choctaws, and Chickasaws into 33 million acres in the southwestern section of the United States. What to do with them became a burning question, inasmuch as they occupied prized territory. Some favored burying them underground; others proposed transporting them beyond the Mississippi River to the Great Plains where they would cease to bother white men for a long time to come—a policy personally favored by Jackson. "Say to [the Creeks]," he wrote one agent, "where they now are, they and my white children are too near to each other to live in harmony and peace. Then game is destroyed, & many of their people will not work, & till the earth. Beyond the great river Mississippi where apart of their nation have gone, their father has provided a country, large enough for them all, and he advises them to remove to it. There, their white brethren will not trouble them . . . and they can live upon it, they and all their children as long as grass grows or water runs in peace and plenty. It will be theirs forever."

The most civilized tribe of Indians in the Southwest were the Cherokees, men of remarkable talents, who lived in fine houses, owned slaves, and cultivated large stretches of fertile soil. They tended to act like members of a separate, sovereign state, capable of running their own affairs. Moreover, they were protected in their rights—or so they thought—by treaties with the United States. But Georgia refused to recognize any special quality about the Cherokees except that they were red men and owned land which the people of Georgia coveted. Consequently, in December, 1829, the Georgia legislature passed a law extending its authority over the Indian territory within the borders of the state, figuring, of course, that the Jackson Administration would support the action. Then, when an Indian named Corn Tassel killed another Cherokee in the territory, he was immediately hustled before a Georgia court, declared guilty, and sentenced to be hanged. The Cherokee Nation appealed to the U. S. Supreme Court on the grounds that Georgia lacked jurisdiction; the court responded by upholding the rights of the Indians against the state, although declaring at the same time that the Cherokees were dependent upon the federal government. Still Georgia disregarded the court, and just a few days before the decision was handed down, executed Corn Tassel.

In another case, one that grew out of a Georgia law forbidding whites to reside among Indians without licenses, several missionaries, two of whom were named Worcester and Butler, appealed to the Supreme Court after their arrest for violating the law. Chief Justice John Marshall again decided against Georgia by decreeing that the Cherokees constituted a definite political community over which the laws of Georgia had no legal force. But again Georgia denied the authority of the court and its sentence. Jackson, on hearing the decision, supposedly remarked, "Well: John Marshall has made his decision: *now let him enforce it.*" It is hardly likely that Jackson was handy with so apt a retort; furthermore, while he sympathized with Georgia's problem and criticized the content of Marshall's decisions, he would never advocate or encourage annulling the federal authority, even that portion of it held by the judicial branch. Nonetheless, the decision remained unenforced because, as the realistic President knew full well, the people in the West and South would not tolerate its enforcement. They wanted the Indians removed—right now. Therefore, in 1830, Congress passed the Indian Removal Act, by which the lands held by Indians within the states were exchanged for new lands west of the Mississippi River. Again, in 1834, Congress enacted the Indian Intercourse Act setting up the Indian Territory where red men might enter and find "perpetual" protection against the white men. A ring of forts would gird the Territory, keeping the red men in and the white men out.

Thus, under Jackson, the tragic removal of the Indian began. During his administration, Old Hickory signed over ninety treaties with various tribes promising them lands in the West, in perpetuity, if they would abandon their tribal homes east of the Mississippi. The horror of the removal beggars the imagination. Often, Indians were tricked into signing away their possessions and then driven off before adequate arrangements could be initiated for an orderly migration westward. They were pressed along "a trail of tears" to find disease, starvation, and death on the western plains. Several times suspicion and distrust precipitated Indian wars. When the hungry Sac and Fox Indians in the north recrossed the Mississippi River to plant grain, the white settlers suspected an attack and proceeded to slaughter the savages as they tried to escape. This was the Black Hawk War of 1832. An even more serious conflict erupted in 1835, when the Seminole Indians in Florida, led by an able young chief, Osceola, refused to comply with the treaty they had signed and evacuate the territory. Although the President sent troops into Florida, the Seminoles were more adept at guerrilla warfare in the Everglades, and the war dragged on for years. Even after Osceola was captured under a flag of truce and died a prisoner at Fort Moultrie in Charleston Harbor, the Indians continued to fight. But their cause was hopeless; though it cost the government nearly fifteen million dollars, the Seminoles were eventually subdued, their lands sequestered, and most of the tribe driven westward.

To one group of Americans, Jackson's removal policy mirrored the national need, that it was a simple response to what the President knew the people demanded as well as a reflection of his own belief, and his western background; to another group, it was one more example of his commitment to States' rights, in this case, the rights of Georgia and other southern and western states to annul the just claims of the Indians. But, as several observant politicians pointed out, Jackson always seemed to be doing two things at once: trying to maintain one foot in the States' rights camp, at the same time he jammed the other foot into the camp of the nationalists. That the President managed this feat so deftly was a testament to his political skill.[2]

But if there was misunderstanding about the President's intention in upholding Georgia's claims, there was no possible confusion over his intentions with respect to South Carolina and the problem of the tariff.

Trouble had been brewing for a long time. Southerners resented the tariff protection accorded northern industries because, among other things, it meant they had to buy their manufactured goods on a closed market, while they sold their cotton abroad on an open one. They were caught in a squeeze that drained them at both ends. Northerners, on the other hand, argued that they had to have government protection if they were to sustain themselves against competition from Europe, particularly Great Britain. Unfortunately the last tariff, which jacked up the rates as high as their manipulators could get them, was conceived solely as a vehicle for Jackson's election, but now that he was safely ensconced in the White House, Southerners expected him to haul the rates down again. Indeed, if he did not, some of them were ready to do it for him, using the nullification device suggested by John C. Calhoun. Acting on the request of James Hamilton, Jr., Calhoun further explained his doctrine of nullification in a statement dated July 26, 1831, which became known as the "Fort Hill Address." According to his view, the Union was a compact of states, in which each state retained the right to examine the acts of Congress, and when necessary nullify within its borders any it felt was a violation of its sovereignty and rights. Thus, Calhoun would assign judicial review to each state, thereby destroying the fundamental concept of the tripartite government constructed under the Constitution. To his credit, he did not favor secession. Nothing so blunt and sudden. What he proposed was the slow but inevitable dissolution of the Union through long legal and constitutional procedures. However, Southerners never really cared for the slow process. They preferred the quick bang. They never lost their affection for secession even when they trifled with nullification.

The testing of these various attitudes and doctrines came with the introduction of a new tariff in Congress by Henry Clay on January 9, 1832. Unfortunately what should have been an honest attempt to eliminate the abominations of the 1828 law degenerated into a political squabble between uncompromising politicians. And, although Van Buren had left for England in September, 1831, the nullifiers were positive he had a hand in the business through his friend, Louis McLane, Secretary of the Treasury—just as he had in 1828. To some extent, they evened the score with him when his nomination as minister to Great Britain was finally reported to the full Senate from the committee on foreign relations. Temporarily setting aside the tariff, the members debated the nomination on January 24 and 25, after which a vote was taken that resulted in a tie, 23 to 23, with the nullifiers joining Clay and the National Republicans against confirmation. The casting vote remained with the Vice-President, John C. Calhoun. Indulging in "exquisite pleasure," he voted to reject the nomination, then he turned to a friend and said, "It will kill him, sir, kill him dead. He will never kick, sir, never kick." But most everyone else had the opposite view: that Van Buren would not only kick, but kick Calhoun right out of his vice-presidential seat.

Predictably, Jackson fumed over the rejection, although Calhoun's spite played right into his hands. Not only did it evoke sympathy for Van Buren, thereby insuring his promotion to higher office, but it drew the Clay and Calhoun forces closer together and rendered it much easier to isolate and defeat them. For example, the President promptly snatched leadership of the tariff question away from Henry Clay—whose bill he mistrusted—by substituting another more to his liking. The southern "nullies" naturally prepared to resist any tariff, unless it completely capitulated to their terms, and just as naturally the President rejected their excessive terms. Instead, he tried to shape a bill which was satisfactory to the North and West without inflicting severe penalties on the South. What resulted was the Tariff in 1832, passed in the House on June 28 by a vote of 132 to 65; and in the Senate on July 9, by a vote of 32 to 16. Generally, the rates of the bill hovered near those imposed by the Tariff of 1824; and although it was not a low tariff, it did include several new items on the free list. Even so, high duties were again imposed on such political essentials as wool, woolens, iron, and hemp. Jackson signed the tariff on July 14, 1832, under the genuine conviction that it represented a compromise, one that would win approval in the North and quell the discontent in the South. Indeed, many Democrats and National Republicans applauded it as a reasonable and judicious "middle course," perhaps not wholly satisfactory to all, but one with which all sections and classes of the country could live.

This tariff reform, approaching sectional balance and accommodation, was totally unacceptable to the South Carolina Nullifiers. Representing none but a small radical fringe, they planned to subvert protection by threatening disunion. Their objective clearly in mind, they organized themselves in South Carolina during the presidential election of 1832 to strengthen their position before applying Calhoun's doctrine of nullification. So effective was their organization that they won a thumping victory at the polls, whereupon the governor of the state, James Hamilton, Jr., summoned a special session of the South Carolina legislature, which in turn called for the election of a convention to meet at Columbia on November 19, 1832, to take appropriate action.

As South Carolina sped toward nullification, possible disunion and civil war, the President countered with several effective actions. First he examined and prepared his military strength. He alerted naval authorities at Norfolk, Virginia, to stand ready to dispatch a squadron to South Carolina if trouble should arise. He ordered the Secretary of the Navy to take appropriate action to check any attempt by the "nullies" to undermine the loyalty of naval officers and men at Charleston Harbor; then he warned the commanders of the forts in that harbor to stand by for a possible

emergency; and finally, he hurried Major-General Winfield
Scott southward to take command of the Charleston garrison that had recently been changed. Yet none of these measures seemed sufficient to cool the hotheads gathered at
Columbia, and on November 24, 1832, the convention, by a
vote of 136 to 26, adopted an Ordinance of Nullification
declaring the tariffs of 1828 and 1832 null and void and forbidding the collection of duties required by these laws
within the state of South Carolina. Also it warned the federal government that if force were used to coerce the state,
South Carolina would secede from the Union.³

Jackson's reaction to this threat was masterful. He did
not, as many contend, respond in a wild outburst of
promises to scourge the state with violence unless it immediately capitulated and obeyed the tariff laws. He proceeded
cautiously, trying to be conciliatory. Nevertheless, he would
not be bullied, nor would he tolerate the humiliation of the
country. What he did was alternate between the gesture of
conciliation and the menace of retaliation. At one moment,
he professed his willingness to forgive and grant concessions; at the next, he let it be known that he was preparing
an army to put down treason. And it was this change of pace
that threw the "nullies" completely off balance, weakening
their resolve to experiment with the most extreme form of
protest.

In implementing his approach, Jackson did several
things at once. He encouraged Unionists within South
Carolina, such as Joel Poinsett, the former minister to
Mexico, assuring them that munitions would be available to
them if it came to a question of saving the Union. "I repeat
to the union men again," he wrote Poinsett, "fear not, *the
union will be preserved* and treason and rebellion promptly
put down, when and where it may shew its monster head."
And, good to his word, he transported arms and ammuni-
tion a short distance away in North Carolina. Also, he sent

George Breathitt, the brother of the governor of Kentucky, to South Carolina, ostensibly as a postal inspector but actually to serve as a liaison with the Unionists and to keep the President informed of developments lest they suddenly deteriorate to the point of secession and require vigorous action. Then, shifting his approach, Jackson delivered to Congress his annual message in which he urged a policy of conciliation. ". . . In justice," he wrote, ". . . the protection afforded by existing laws to any branches of the national industry should not exceed what may be necessary to counteract the regulations of foreign nations and to secure a supply of those articles of manufacture essential to the national independence and safety in time of war. If upon investigation it shall be found, as it is believed it will be, that the legislative protection granted to any particular interest is greater than is indispensably requisite for these objects, I recommend that it be gradually diminished, and that . . . the whole scheme of duties be reduced. . . ." Jackson ended by noticing the danger in South Carolina but ventured that the laws were sufficient to handle any eventuality. Privately, he admitted something more was necessary. "The union . . . will now be tested by the support I get by the people," he wisely said. "I will die with the union."

Then, in a Proclamation dated December 10, President Jackson spoke directly to the people of South Carolina. In his message, he blended words of warning with entreaty, demand with understanding, threat with conciliation. He appealed to their fears, their pride, their interests; at the same time he categorically rejected nullification and secession. The nation was supreme, he said; not the states. "The laws of the United States must be executed. I have no discretionary power on the subject; my duty is emphatically pronounced in the Constitution. Those who told you that you might peaceably prevent their execution deceived you. . . . Their object is disunion. But be not deceived by names. Disunion by armed force is *treason*. Are you really ready to incur its guilt?"

Much of the message was the work of the Secretary of State, Edward Livingston, particularly in its strong constitutional arguments about the nature of the Union, arguments very similar to those advanced by John Marshall and Daniel Webster. But the fire in it was pure Jackson; and that fire was what gave the Proclamation its special strength and power.[4]

The publication of the Proclamation produced a chorus of patriotic shouts around the country. Meetings were organized to express support of the President; parades and bonfires demonstrated the ardor of Americans to stand behind Jackson. Thus, the Proclamation not only rallied the people to the President's side, prompting state legislatures (including those in the South) to denounce nullification and assure the General of their loyalty, but it also warned the nullifiers that if they rejected a peaceful settlement he was quite prepared to summon an armed force to execute the laws. In this respect, Jackson quietly let it be known that he could have fifty thousand men inside South Carolina within forty days and another fifty thousand forty days after that. Then, in a public display of his intention, he asked Congress for the necessary legislation to permit him to insure obedience of the tariff laws in South Carolina and the collection of the custom duties. When introduced into Congress the Force Bill (or Bloody Bill as it was called by some) received widespread support on a nonpartisan and intersectional basis.

Precisely within a week after Jackson's annual message, South Carolina indicated a willingness to respond to reason. Indeed, even when the Ordinance of Nullification was first adopted, it was apparent that a negotiated settlement was possible by the very fact that the date for the Ordinance to go into effect was advanced to February 1, 1833, giving the national government enough time to suggest an agreeable compromise. On December 10, the South Carolina legislature went further and elected as governor, Robert Y. Hayne, who was much more moderate on the nullification issue than his predecessor. Two days later, it elected John C. Calhoun U.S. Senator, thereby stationing the state's strongest bargainer inside the Congress to work out a settlement. Calhoun resigned the vice-presidency on December 28, 1832, to take his new office and begin the task of finding a consensus that would spare his state the humiliation of military defeat. As he was sworn in, one observer said, "I could not help thinking when he took the oath to support the Constitution of the United States that he made a mental reservation that it would be as 'he understood it.'"

Gratified by South Carolina's seeming efforts at moderation, Jackson quickly responded to show the state that any favorable move to avoid violence would be met with similar forbearance. For example, although General Scott had been sent to take command of the South Carolina troops, he was repeatedly warned to avoid trouble and any unnecessary display of force. Meanwhile, the administration threw its full weight behind a new tariff bill, introduced by Gulian C. Verplanck of New York and written with the assistance of the Secretary of the Treasury, which would lower protection by 50 per cent within two years. The bill elicited strong support from many northerners worried about the possibility of secession and civil war. Unfortunately, protectionists howled their fears over such a sharp reduction of duties proposed by the Verplanck Bill. To make matters worse, Henry Clay tried to steer a land bill through Congress that would distribute revenue from the sale of land to the states, thus reducing the government's revenue and forcing it to raise the tariff. It was "mischievous" business, to use Silas Wright's apt description of it, completely unworthy of its author, but Clay had just been dealt a decisive defeat by Jackson in the presidential election of 1832, and he was bitter, angry, and resentful over his loss of prestige. Despite his nationalistic sentiments, he refused to assist the President in his efforts at compromise. Instead, he plotted to scuttle any legitimate accommodation of the tariff question. He seemed more worried by Jackson's triumphs and popularity than South Carolina's defiance of law. To strengthen his tactical position in Congress, therefore, he concluded an alliance with Senator Calhoun on the assumption that such a coalition would operate to their mutual benefit and the President's discomfort. Working together, Clay and Calhoun would kill the Verplanck Bill and emasculate the Force Bill.

On February 12, 1833, Clay introduced into the Senate what was euphemistically called a "compromise" tariff. The bill provided for the reduction of rates over a ten-year period, at the end of which time, no duty would be higher than 20 per cent. But there was a joker buried in it: only the tiniest reductions would occur during the first years of the bill's operation, the major changes coming nine years later at the tail end of the period. And, as it turned out, they would prove so sharp a drop as to threaten the economy of the country. It was a bad law, written without regard to compromise, and done for political advantage and the caressing of Clay's *amour propre*. However, if it could extinguish the fires of nullification and disunion the administration would go along with it, however apparent the bill's defects. The measure passed the Senate on March 1 by a vote of 29 to 16, and the House on February 26 by a vote of 119 to 85. Almost all the South voted to accept it—the Congressmen literally falling over themselves to agree to any solution that would end the dispute and the possibility of civil war. New England and some of the Middle Atlantic states tended to vote against the bill, while the Northwest was split. That the passage of the Compromise Tariff of 1833 should be due almost entirely to the massive support it received from the South proved to some men how ridiculous politics could be.

While the tariff hared its way through Congress, the Force Bill was brought up for debate. Calhoun denounced the measure, finding it a suitable occasion to reargue his interpretation of the nature of the Union. He made a final effort to kill the bill by forcing an adjournment, but the Congress, conscious of the meaning of compromise, vetoed this maneuver. When the vote on the Force Bill was taken in the Senate, Calhoun and his followers strode out of the chamber. Only John Tyler of Virginia remained in his seat to cast the single vote against the bill. Reluctantly, Clay voiced approval of the measure during the debate, but on the day of the final vote he failed to appear, claiming poor health and the need to stay home and rest. The bill passed the House on March 1 by a vote of 149 to 48.

With great pride, Jackson signed both the Compromise Tariff and the Force Bill on March 2, 1833, just as his first term in office ended. South Carolina reassembled its convention on March 11 and repealed its nullification of the tariff laws but then proceeded to nullify the Force Bill. It was a pathetic gesture to save face, and the President chose not to quarrel with that.

Jackson's victory was an extraordinary display of tact and rare wisdom. He did not gain a total victory over South Carolina, nor did he want one. In politics total victory usually means eventual defeat. What he did was spare the Union the agony of civil conflict. He accomplished it not by waging war but by initiating compromise. That he did everything in his power to oblige South Carolina can be seen in the remark of John Quincy Adams, who thought the controversy was won by the Nullifiers, not Jackson. Of course, Calhoun and his henchmen claimed victory, and some even thought the real victor was Henry Clay, who, through it all, had managed to preserve his precious tariff. But Jackson's accomplishment can not be dismissed so lightly. Through the careful use of presidential powers, ral-

lying the people to his side, alerting the military, offering
compromise while preparing for treason, he preserved the
Union and upheld the supremacy of federal law.[5]

Because he had grown tremendously in political sagacity
during the past few years Jackson took off on a long tour of
the country just as soon as he could escape Washington. His
purpose was to encourage demonstrations of popular
approval for his recent conduct in handling South Carolina.
The trip enabled the public to see their President close up as
well as express their "detestation of nullification." Leaving
the capital in late spring, 1833, and accompanied by Van
Buren, Levi Woodbury, Lewis Cass, Major Donelson and others, Jackson journeyed to Baltimore, Philadelphia, Trenton,
New York, Boston, Newport, Salem and many other cities.
Those who witnessed any part of the grand tour never forgot
"the long processions; the crowded roofs and windows; the
thundering salutes of artillery; steamboats gay with a thousand flags and streamers; the erect, gray-headed old man,
sitting his horse like a centaur, and bowing to the wild hur-
rahs of the Unterrified with matchless grace; the rushing forward of interminable crowds to shake the President's hand;
the banquets, public and private; the toasts, addresses,
responses; and all the other items of the price which a popular hero has to pay for his popularity."

In New York City, Jackson took a steamboat to Staten
Island and while staring at the city's magnificent harbor he
suddenly turned to a companion and blurted out, "What a
country God has given us! . . . We have the best country and
the best institutions in the world. No people have so much
to be grateful for as we."

Further along in his tour, at Cambridge, Massachusetts,
the President received an honorary degree of Doctor of
Laws from Harvard—much to John Quincy Adams' intense
disgust. Later, when Jackson addressed a small audience in
another town someone called out from the crowd, "You

must give them a little Latin, *Doctor.*" Whereupon the President, his eyes twinkling, responded: "*E pluribus unum,* my friends, *sine qua non!*"

Jackson was forced to cut short the tour at Concord because of extreme fatigue and "bleeding at the lungs," but by that time he had succeeded in his primary mission of summoning popular support for a strong federal government in its quarrel with a rebellious state. Here was Jackson at his political best, using his office and popularity to unite the American people.

Indeed, it was Jackson's profound understanding of his presidential powers and the extent to which he used them that reveals his greatness as a President. For it was Jackson who first explored the full dynamic potential of the American government. That potential largely depends on the initiative and aggressiveness of the chief executive, and none of the previous presidents had sought this kind of leadership while in office. Jackson was different. One example of his creative aggressiveness was his exercise of the veto. Since the founding of the Republic under the Constitution the veto had been used nine times, only three of which dealt with important legislation. In effect, it had been reserved for the voiding of flagrantly unconstitutional measures. Nothing more. Under Jackson, it became something else again: not a negative instrument but a positive weapon to implement the executive purpose. This was done principally by disapproving legislation for reasons that had nothing to do with their constitutionality. The veto thus became both a persuader to encourage Congressional respect for the President's wishes, and a club to chastise those who incurred his displeasure or tampered with his legislative program. In his second annual address to Congress, Jackson assumed "the undoubted right . . . to withhold his assent from bills on other grounds than their constitutionality." It was a bold and forth-right assertion of presidential prerogative and, in the words of one historian, a successful claim that the weight of the chief executive was the equivalent of two-thirds of both houses of Congress.

Jackson also employed the pocket veto for the first time. This is a special veto, by which the President may kill legislation by withholding his signature from a bill after the Congress has adjourned. One such pocket veto was delivered in 1833 against Clay's "mischievous" land bill.

In the course of his two administrations, Jackson vetoed twelve bills, more than all of his predecessors combined. And sometimes when he vetoed his messages trumpeted with appeals to the people, alerting them to the foolishness that had just skidded through Congress. More than any other president before him, Jackson used his office to reach the people, employing messages, Proclamations, Protests, and any number of party devices, such as newspapers, to close the distance between the chief executive and the American electorate. He also used his office to reach into Congress and control legislation. Not only did his advisers work closely with Democratic leaders in both houses, but he himself repeatedly sought to control the membership of Congressional committees.

But in the assertion of his presidential rights, Jackson did not simply blurt them out and invite the country to accept them or go to the devil. His tactics, bold in design, were slow in execution. Again, his use of the veto provides a suitable example. The Maysville Road bill was his first veto, and in it he presented the strictest kind of constitutional argument. It was the sort of thing the people could be expected to accept, however disappointed they might be about the loss of the road. After that, he moved nearer his goal by voiding certain bills because of what he called their "impropriety," without clearly indicating what he meant by "impropriety," whether they were unconstitutional or legally deficient somehow or what. Not until he had been in office almost two years did he announce the presidential privilege to disallow legislation on grounds other than their constitutionality. And when he finally did so, it was with great "reluctance and anxiety," he said, as though not wishing to make the claim but that he had no choice. Once he had gotten that far the rest was easy. Thereafter, when he struck down major legislation, he would assert all manner of reasons: political, economic, social—anything that he regarded as contrary to the public interest.

Thus, with Jackson begins the history of dynamic and aggressive executive leadership in the United States. Those in his own generation and later who, for one reason or another, insisted on characterizing him as an arrogant, militaristic hothead, slamming around the White House and shooting from the hip and lip at the slightest provocation, will never admit to his statesmanship or understand his contribution to the presidency. But those who will gage his skills and insights into the political process, and measure the distance he stretched the executive powers, will discover some of the factors that constitute his greatness as an American President.[6]

CHAPTER VIII

The Bank War

It was the heroic struggle with the Second Bank of the United States that truly revealed Jackson's conception of presidential powers. Moreover, it revealed his own special brand of democracy. Pursuing a calculated course, he proceeded to destroy the Bank because he believed it capable of infinite harm to the American people and government. It was, in his own celebrated words, a "monster," a "hydra-headed monster," whose powers and potential were so enormous that it threatened the safety of the Republic.

Andrew Jackson destroyed the Bank, no one else. Not Martin Van Buren, not Wall Street bankers, not the Albany Regency, Amos Kendall, Roger Taney, or anyone else. Granted Jackson had the assistance of a number of shrewd, dedicated, and conscientious political operators, notably Kendall, Taney, Benton, Francis P. Blair and others, but they were all of secondary importance. It was Jackson who conceived the idea to smash the financial giant, and it was Jackson who determined how and when it should be done.

What was his motive? To some extent, the President was conditioned by his financial losses at an earlier time, possibly his views on States' rights and maybe even western prejudice, but fundamentally the Bank War was not an economic struggle, nor a contest between sections, nor an expression of hostility by the frontier toward the city, nor an act of spite or revenge; essentially it was a political war, or at least that is how it started. After 1833, it developed into something else again. But, in initiating the battle, Jackson's motivation was fundamentally political. He regarded the

[163]

Bank as dangerous to the liberty of the American people because it represented a fantastic centralization of economic and political power under private control. It was a "monopoly" with special privileges, and yet it was not subject to presidential, congressional, or popular regulation. Only the financial interest of the B.U.S. constituted any real control of this "monster," and to Jackson that was no control at all.

The political motive, therefore, is absolutely central to the Bank War, at least in its initial stage. However, it was not a simple case of Jackson battling for democracy against a moneyed aristocracy, the few against the many, the rich against the poor, although obviously these can and have been read into the controversy. What Jackson wished to terminate, what he hammered at over and over again, was the Bank's enormous political "power to control the Government and change its character. . . ." He blasted the institution as "an irresponsible power" spending its money "as a means of operating upon public opinion." He termed it a "vast electioneering engine."

For the President to decide by himself that this brain child of Alexander Hamilton had no right to exist in its present form took towering presumption. But Jackson was never especially deficient in that department. It also took courage—another of his virtues. Finally it required extraordinary political skill to bring off the complete destruction of the Bank, and here is where the President really excelled. Yet, despite a deep-seated prejudice against all banks that went back many years, he was very slow to move against the biggest of them, possibly because of his willingness to compromise with it at first, possibly because of his habitual caution, and possibly out of a need to instruct the people, his friends (with some notable exceptions), and Congress at almost every step of the way.

Following the initial battle with the Bank, when recharter of the institution was denied, the war intensified and developed into a disordered contest among politicians and businessmen from every section of the country who scrambled for advantage at the Bank's expense—everyone for himself, pushing for the main chance. It was an incredible demonstration of lusty men in a swiftly changing age struggling to get ahead by annihilating whatever stood in their way.

In other words the Bank War involved two phases: the first, stretching from 1829 to 1833, constituted a power struggle between Jackson and the Bank over the President's fear of the Bank's unchecked political and economic privileges; the second, continuing after 1833, was a highly complicated story of political and economic jockeying among "men on the make" who were out for anything they could get, each man for himself, snatching at every advantage.[1]

The Second Bank of the United States was chartered by Congress in 1816, with the charter to run twenty years. The capital stock of the B.U.S. was assigned at 35 million dollars, one-fifth of which was subscribed by the federal government, the rest by the public. A board of twenty-five men directed the Bank's affairs, appointed its administrative officers, invested its funds, and established its branch banks in the principal cities throughout the country. Five of the board members were appointed by the President of the United States, the others elected by the stockholders. In effect, the B.U.S. was a central bank, designed to regulate the credit and currency operations of the country. This it did, and did extremely well, despite Jackson's claims to the contrary. The Bank had authority to issue notes that were receivable for money owed the U.S. government, and it served as a depository for government funds that could be invested for the profit of the stockholders. Thus, the B.U.S.

exercised tremendous influence in foreign and domestic finance, as well as unrivaled power over state banks. The profits from this tidy arrangement were shared by private investors, both American and foreign, and by the federal government.

The first president of the Second Bank was William Jones, a man of rare incompetence, a onetime Secretary of the Navy and Secretary *pro tem* of the Treasury. He was followed in 1819 by Langdon Cheves of South Carolina, a former Congressman and Speaker of the House of Representatives. The same year the presidency changed hands an economic depression struck the country. The panic was part of a world-wide dislocation, but it was intensified in the United States by Cheves' policy of restoring the Bank's credit by calling in loans and foreclosing mortgages. The B.U.S. gathered the notes of many state banks and then returned them for payment in cash: gold and silver. Since many of these unlucky banks were without specie, they slid into bankruptcy. The resulting depression brought a price collapse, unemployment, and, in some areas, starvation. It was especially severe in the West where political repercussions continued to echo for almost ten years.

But the B.U.S. was saved. It had gone through a terrible ordeal under Jones' mismanagement, but Cheves had steered it back to a position of financial strength and soundness. His work done, Cheves retired and was replaced as president in January, 1823, by Nicholas Biddle, a thirty-seven-year-old scion of a wealthy and distinguished Philadelphia family. Biddle, himself, was one of those impossibly gifted men. He had everything: brains, looks, money and family, and now he assumed control of a financial colossus. Although he exercised this control with considerable restraint and discretion, nevertheless at times he could be arrogant and would prod the monster to a vicious display of its power. The Bank was subject to no regulatory check except what was imposed by the laws of business and the profit-minded demands of the stockholders. Under Biddle it prospered, branching into twenty-nine cities from its headquarters on Chestnut Street in Philadelphia and doing a business of 70 million dollars a year. As a central banking system, it provided the financial operations of the country with uniformity and regulation, and as such was of immense economic benefit to all the people. Despite rumors to the contrary, it handled only 20 per cent of the country's loans; its note circulation was only one-fifth of the nation's total; and it held only one-third of the total bank deposits and specie.

Even so, to Andrew Jackson it was a monster, maybe half asleep, but hardly less dangerous. It was not so much the size of the beast that offended, though that was bad enough, or the amount of money stored in its care, but rather that it was unchained, that it was independent of the government and the people, that it had the means to nullify economic development, and that it frequently interfered in politics. A few years later, in summing up the administration's complaints against the institution, Roger B. Taney listed "its corrupting influence . . . its patronage greater than that of the Government—its power to embarrass the operations of the Government—& to influence elections. . . ." These were all basically political reasons, and the last-named—"influence elections"—was the one that especially concerned Jackson. It was really impossible for him to believe that a Bank with so much money and so many special privileges could remain independent of the political process. And such involvement, said Jackson, threatened "to destroy our republican institutions." In June, 1829, just three months after his inauguration, the President told John Overton that he planned to change "the present incorporated Bank to that of a National Bank—This being the only way that a recharter to the present U.S. Bank can be prevented &

which I believe is the only thing that can prevent our liberties to be crushed by the Bank & its influence,—for I [have learned] of the injurious effect & interference of the directors of the Bank had in our late election which if not *curbed* must destroy the purity of the right of suffrage."

Another thing: the monster discriminated. It did not respond when Andrew Jackson, the duly elected and lawful head of the national government, commanded; it only responded when Nicholas Biddle, the representative of a select group of stockholders, gave the order. It discriminated in other ways, too. It extended favors to a few men for the help they could render the Bank; and some of these men were important politicians, such as Daniel Webster, Henry Clay, William T. Barry and others, men on both sides of the political aisle, and who were indebted to the Bank for many thousands of dollars. This was another way it interfered in politics. At one point, Webster wrote to Biddle, "I believe my retainer has not been renewed or *refreshed* as usual. If it be wished that my relation to the Bank should be continued, it may be well to send me the usual retainer. . . ." Now Webster served as legal counsel to the B.U.S., but his "retainer" covered more than his legal fees though a lot less than the "corrupt" intent he was accused of by Democrats.

That the Bank, therefore, did not generally abuse its privileges and that, by 1830, it contributed substantially to the welfare of the country is clear. That at any moment and for any reason it could also abandon its responsibilities and scorch the nation with its fiery breath as it had done in 1819 is also very clear.[2]

Jackson came to the presidency in March, 1829, with every intention of "chaining" the monster. His prejudice—no doubt nurtured by the bone-rattling financial misadventures of his early days and especially the harrowing history of the Allison land deal which forever confirmed his hostility to paper money and the agencies that issued it—was encouraged by several of his advisers. They assured him that the Bank's vast political power had been used against him in the presidential election of 1828, and that it would continue to be used against him in the future. Amos Kendall, for example, whose own hatred for the Bank emerged during the Relief War in Kentucky five years before, and Isaac Hill, who presented the President with documentary proof that the branch of the Bank in New Hampshire discriminated against friends of the administration, were especially effective in this regard. But, as one member of the Cabinet pointed out, Jackson had expressed "strong opinions against the Bank of the United States" even before becoming President.

In his first message to Congress in December, 1829, the Hero alerted the B.U.S. to his intentions. Those two short paragraphs at the end of the message were the tip-off. Interestingly enough, in this first warning, Jackson did not propose the destruction of the Bank. Probably, at first, he did not intend to kill it outright. He was willing to compromise with it on the basis of certain changes that would eliminate his constitutional objections. Also, he needed time to convince the people, his party, and Congress that tampering with the Bank was not the deranged suggestion of a lunatic westerner.

But Biddle would never compromise. Why should he? He was doing a splendid job, had many supporters in and out of Congress, and enjoyed the knowledge that the country needed and wanted his Bank—just as it was. But, to play it safe, he insinuated his way into the Jackson circle of friends by doing financial favors for those who had the good sense to appreciate them, not realizing that he was proving Jackson's contention that the Bank played politics. As added protection, Biddle generously proposed to pay off the national debt by January 8, 1833—the eighteenth anniversary of Jackson's victory over the British at New Orleans—

recognizing how appreciative the President would be for liquidating the debt and for doing it on such a memorable day. Of course for this gracious gesture, there was a price: the continuation of the Bank's charter.[3]

Whether he realized it at the time or not, Jackson had touched a very tender spot in the breast of many Americans by his message, particularly those seeking economic advantage who had been denied the assistance of the B.U.S. Several wrote the President complaining how the Bank was contemptuous of their demands and favored only those whose operations guaranteed huge profits to fatten the dividends of the great stockholders. Other would-be antagonists of the Bank were roused by Jackson's message: state bankers, for example, especially in New York and Baltimore, who resented the size, wealth, and controlling functions of the B.U.S. and the amount of profits pumped into Philadelphia; also, freeholding farmers, who regarded the Bank as the embodiment of all the corrupting forces in society threatening their simple republican way of life; and urban wage earners, who viewed the Bank as the largest monopoly among many monopolies equipped with special privileges to grind the faces of the poor; finally, there were lawyers and other professional men, small planters, merchants, and manufacturers—a diverse group of Americans, having at least one thing in common: they were all economically aggressive.

Yet despite this broad spectrum of sympathy for his apparent intention to modify the Bank's charter, Jackson did not proceed any further. He bided his time, for he was not a man to act precipitously. He talked about possible changes in the operation of the Bank without indicating any urgency in his suggestion. For instance, at one time he tossed out the idea of tying the B.U.S. to the Treasury and restricting its note-issuing powers; at another, he thought of recommending to Congress the creation of a government-owned institution with branches in the several states. Otherwise, he did nothing. Unquestionably, part of his inaction during the early years of his administration was due to his problems over the Eaton affair and the breakup of his first Cabinet. Yet, when he chose his new Cabinet in 1831, he selected men who favored a National Bank—or some variation of it. They were not expressly hostile to a continuation of Biddle's institution, and probably most of them doubted that the matter would come up for a number of years, since the charter of the B.U.S. had several years to go before expiring in 1836.

But they figured without Biddle. Suddenly, almost without warning, he decided to ask for a recharter in 1832, four years before the date of expiration. On the face of it, this action seemed an excellent idea, since 1832 was a presidential election year. Biddle guessed that Jackson would never risk re-election by making the Bank an issue in the campaign. Inasmuch as he must request recharter anyway, Biddle felt he had a better chance of getting it if he asked before the election rather than afterward. Besides, if Jackson refused the request and vetoed the recharter bill, then Congressional candidates up for election would have to commit themselves on the issue one way or the other, and Biddle believed there were enough Americans who favored the B.U.S., and that a sufficient number of Congressmen would be elected to override the veto. Henry Clay, John Quincy Adams, and Daniel Webster concurred in the wisdom of the course. But what they failed to tell Biddle was that the action deliberately baited Jackson—it goaded him into a veto.

"Now as I understand the application at the present time," wrote Roger Taney, "it means in plain English this— the Bank says to the President, your next election is at hand—if you charter us well—if not—beware of your power." This was it precisely. Other men who were friends of the Bank and friends of the administration warned Biddle that the application would resoundingly prove the President's contention that the Bank was a political agency interfering in the electoral process and was in effect ordering the government either to comply with its will or be prepared to sustain a "severe chastisement."[4]

Clay disagreed with this argument, however. "The course of the President in the event of the passage of the bill," he told Biddle, "seems to be a matter of doubt and speculation. My own belief is that, if *now* called upon he would not negative the bill. . . ." Biddle concurred, and he initiated action by writing to Senator George M. Dallas of Pennsylvania on January 6, 1832, and presenting a memorial asking for renewal of the Bank charter. Three days later, the memorial was placed before each house of Congress, and, in turn, was referred to separate committees: one in the House chaired by George McDuffie and one in the Senate presided over by Dallas.

Two months later, the Senate bill came to the floor for debate, and the support for the measure was so strong and so spontaneous that immediate action was necessary if the President's objections were to get a proper hearing. After a consultation among Democratic leaders, it was decided to move a House investigation of the B.U.S., thereby giving anti-Bank Jacksonians time to consolidate their forces and perhaps come up with information to discredit Biddle's company in the eyes of the American people. In the forefront of this fight in Congress stood Senator Thomas Hart Benton of Missouri, "Old Bullion," whose views about the danger of paper money corresponded exactly with those of the President. In the ensuing battle, he was an army in himself, converting Democrats into anti-Bank men, drilling Congressional forces against Biddle, and through his speeches in the Senate propagandizing the country about the evils practiced by the monopoly. He even arranged the details for the investigation of the Bank initiated in the House of Representatives, and he saw to it that the eventual report by the majority of the committee was loaded with enough examples of alleged abuses by the Bank to substantiate the President's case against recharter.

Though the subsequent report shimmered with distortion, it made beautiful propaganda. Its charges were restated in Democratic newspapers across the country, libeling the Bank with claims of having violated the terms of its charter. The Washington *Globe,* edited by Francis Blair, led the other party organs in circulating the document and in arousing public opinion in defense of the President's cause. It launched a massive attack to "display the evil of the [B.U.S.], rouse the *people* [and thereby] prepare them to sustain the veto." It was a magnificently conducted campaign at all levels, one that again demonstrated the organizational splendor of the Democratic party.

Despite the strategy directed by Benton in Congress, the pro-Bank forces in Congress led by Clay and Webster could not be turned. Even some Democrats went along, fearing that without recharter "the country will be ruined . . . & that there will be no sound currency extant." On June 11, 1832, the bill for recharter passed the Senate by a vote of 28 to 20, and almost a month later, on July 3, it passed the House by 107 to 85. The vote indicated solid support for the Bank in New England and the Middle Atlantic states, strong opposition in the South, and an almost divided opinion in the Northwest and Southwest. Biddle was ecstatic over the vote. "I congratulate our friends most cordially upon this most satisfactory result. Now for the President. My belief is that the President will veto the bill though that is not generally known or believed."

Ill, tired, and debilitated by the hot, sticky Washington weather, Jackson was irritable and cranky when the recharter bill landed on his desk for signature. Martin Van Buren, just returned from England after his rejection by the Senate, came to see the President and found him lying on a sofa in the White House, looking more like a ghost than a man. On seeing his friend for the first time in many months Jackson brightened, then reached out and grasped the Magician's hand. "The bank, Mr. Van Buren is trying to kill me," he said, *"but I will kill it!"* His voice, reported the Red Fox, was entirely devoid of passion, anger, or bluster, but there was no mistaking his mind or mood. He was indeed determined to kill the Bank, not compromise with it, not change its charter and nudge it into the orbit of governmental control, but eliminate it once and for all. What convinced him of this was undoubtedly Biddle's action in seeking recharter at a time, carefully selected, whereby he thought he could tamper with the electoral process to get what he wanted.

Yet, Jackson was too good a politician to miss the danger to his re-election and the election of other Democrats if he vetoed the charter. Politically speaking the moment for battle was all wrong. An election year is no time to disturb the public mind with momentous, quarrelsome issues—not when they are unnecessary and can be avoided. The public is never anxious to settle portentous questions, least of all in an election and certainly not when members of their own party are themselves divided over the issues. The voters bitterly resent being stirred up and having a choice thrust upon them; it agitates and frightens them; and Jackson, by playing into Biddle's hands, was unduly exciting the people and thereby courting electoral disaster. As professional a politician as Silas Wright of New York understood this and told one friend that if the Democrats lost the election it would not be for the want of organization or spirit but because the people "were not equal to the conflict." Still Jackson believed he was acting for the public good in eliminating this extragovernmental "power in the State," that the people would recognize this and that they would sustain him. "Providence has had a hand in bringing forward the subject at this time," he told Kendall, "to preserve the republic from its thraldome and corrupting influence." Thus, certain of popular approval, and convinced of the Bank's evil influence, Jackson made up his mind to write the veto.[5]

The actual writing of the message was the work of several men: Jackson himself, Amos Kendall, Roger B. Taney, Andrew J. Donelson, and Levi Woodbury. What the President wanted was a message that had force and logic and strength to carry it across the nation and convince the people of its fundamental truth. Naturally he needed a closely reasoned paper, but he also desired one that would stir men and reach their minds and hearts, one that could later serve as a propaganda document during the election. His assistants, working three days at fever pitch, did not fail him. What they produced was the most important presidential veto in American history, a powerful and dramatic polemic that can still reach across a century and more of time and excite controversy among those who study it.

In the veto, delivered to the Congress on July 10, the President claimed that the Second Bank of the United States enjoyed exclusive privileges that, for all intents and purposes, gave it a monopoly over foreign and domestic exchange. Worse, some eight millions of the Bank's stock was held by foreigners. "By this act the American Republic proposes virtually to make them a present of some millions of dollars," said Jackson—and why should the few, particularly the foreign few, enjoy the special favor of this country. "If our Government," he continued, "must sell monopolies . . . it is but justice and good policy . . . to confine our favors to our own fellow citizens, and let each in his turn enjoy an opportunity to profit by our bounty." Over and over, like the intense nationalist he was, Jackson repeated this foreign-influence theme, no doubt striking fire in the hearts of

millions of patriotic Americans. Then he turned to the constitutional question involved in recharter. He noted that the
Supreme Court in the case McCulloch *vs.* Maryland had
judged the Bank constitutional. "To this conclusion I can
not assent," announced Jackson. Elaborating, he declared
that the Congress and the President as well as the Court
"must each for itself be guided by its own opinion of the
Constitution. It is as much the duty of the House of
Representatives, of the Senate, and of the President to
decide upon the constitutionality of any bill or resolution
which may be presented to them for passage or approval as
it is of the supreme judges when it may be brought before
them for judicial decision. The opinion of the judges has no
more authority over Congress than the opinion of Congress
has over the judges, and on that point the President is independent of both. The authority of the Supreme Court must
not, therefore, be permitted to control the Congress or the
Executive when acting in their legislative capacities, but
to have only such influence as the force of their reasoning
may deserve." This is an extraordinary concept. In effect
what Jackson said was that no member of the tripartite
government can escape his responsibility to consider the
constitutionality of all bills and act as his knowledge and
good judgment dictate. And in this instance, Jackson did
not agree with the Court about the Bank. Since the matter
was subject to legislative and executive action, he simply
claimed the right to think and act as an independent member of the government.

However, it was toward the end of the message that he let
fly his verbal thunderbolts. "It is to be regretted that the rich
and powerful too often bend the acts of government to their
selfish purposes," was the opening shot. "Distinctions in
society will always exist under every just government.
Equality of talents, of education, or of wealth can not be
produced by human institutions. In the full enjoyment of

the gifts of Heaven and the fruits of superior industry, economy, and virtue, every man is equally entitled to protection by law; but when the laws undertake to add to these natural and just advantages artificial distinctions, to grant titles, gratuities, and exclusive privileges, to make the rich richer and the potent more powerful, the humble members of society—the farmers, mechanics, and laborers—who have neither the time nor the means of securing like favors to themselves, have a right to complain of the injustice of their Government. There are no necessary evils in government. Its evils exist only in its abuses. If it would confine itself to equal protection, and, as Heaven does its rains, shower its favors alike on the high and the low, the rich and the poor, it would be an unqualified blessing. In the act before me there seems to be a wide and unnecessary departure from these just principles."

The passage, proclaiming the doctrine of government as honest broker, is a throbbing exposition of the social philosophy of Jacksonian Democracy. It is bold and aggressive, dramatic and compelling. Jackson's friends rushed forward to declare their enthusiasm for it, pronouncing it—of all incredible things—a second Declaration of Independence. The National Republicans naturally lambasted it, and Biddle likened it to "the fury of a chained panther biting the bars of his cage"; it was a "manifesto of anarchy," he cried, "such as Marat or Robespierre might have issued to the mobs" during the French Revolution. Indeed, it was a manifesto, slightly demagogic in tone, summoning the masses to the President's side. The veto beckoned the nationalist, the farmer, the mechanic and the tradesman; it called to hard money men who disliked banks for their unsound paper money as well as soft money men who criticized the B.U.S. for restricting inflation; and it generally aroused the great numbers of "men on the make" who opposed all monopolistic institutions that hobbled their ambitious reach for economic advancement.

The veto was a hardheaded and realistic political document, full of the sounds that stir men to battle. As an economic document, indifferent to the Bank's genuine services to the nation, it can be easily faulted; as political propaganda it is a masterpiece.[6]

And it was upon the political field in a presidential contest that the Democrats now tested the veto and Jackson's gamble with his popularity. Congress was unable to override the veto, and thus the Bank issue was placed before the American people for decision in the presidential election of 1832. Biddle poured money, propaganda and Henry Clay into the campaign for recharter. But the Democrats were smarter. They had some regard for the delicate sensibilities of the American people, so they screened the unpleasant issue behind a campaign of parades, songs, illuminations, hickory pole raisings, barbecues, and—most of all—the charismatic appeal of that "man of the people," that symbol of democracy, General Andrew Jackson, the Hero of New Orleans.

As it developed, the election of 1832 became a three-cornered contest. Since the disappearance of William Morgan, the Anti-Masonic movement had spread from New York into Vermont, Connecticut, Massachusetts, Rhode Island, Pennsylvania, Maryland, Ohio, and Indiana. Now, because Henry Clay, the likely nominee of the National Republicans, was a Mason, and because Jackson was a "grand king" of that order, the Anti-Masons formed, the first third party in American history. Furthermore, they held the first national nominating convention, meeting at Baltimore on September 26, 1831. In all, 116 delegates representing 13 states assembled and nominated William Wirt of Maryland (who had once been a Mason but had since resigned) for president and Amos Ellmaker of Pennsylvania for vice-president. A few months later, on December 12, 1831, the National Republican party held its convention in Baltimore, and the 156 attending delegates, representing 17 states, nominated Henry Clay of Kentucky and John Sergeant of Pennsylvania.

The Democrats convened last. Three hundred and thirty-four delegates from every state but Missouri (whose delegates could not reach Baltimore in time) met on May 21, 1832, in a public demonstration of party unity and harmony. There was no question about the presidential nomination. The only real business was the vice-presidency that Jackson was determined to present to Van Buren for his loyal service and as recompense for the shabby refusal of the Senate to confirm his nomination as British minister. Establishing the two-thirds vote-requirement for nomination as another sign of their solidarity, the Democrats proceeded to nominate the Magician on the first ballot, giving him 208 votes. After that the convention adopted a unanimous resolution approving the choice of Van Buren, followed by a second resolution concurring in the nominations for president that Jackson had already received from many states. Interestingly enough, this was the only reference to Jackson during the entire convention.

In the ensuing campaign, Henry Clay eagerly grappled with Jackson over the recharter of the Bank, for he rightly surmised that it was the only issue by which he might throw the President to defeat. In previous local elections where the choice of Jackson or the Bank was clearly delineated, Clay observed that the people voted for the Bank because they feared such financial consequences as occurred in 1819. If this action could be raised to the national level, the President could be toppled from office. It was an ambition that Clay had nurtured for years.[7]

Meanwhile, the Democrats glibly tongued all the clichés about the rich and the poor, the plutocrat and the working-man, the few and the many. Their campaign rhetoric fulminated against the moneyed monopoly in the country that, they insisted, was manipulated by an aristocracy intent on frustrating the economic progress of the ordinary citizen. It was a joyful outburst of noise and nonsense, and so extreme in tone and character as to be patently ridiculous; nevertheless, it was just the sort of thing to delight party hacks and reassure a frightened electorate that "cuckoo-politics" were still fashionable; for, as long as politicians can talk in this absurd way, most people figure the country to be out of immediate danger. As for the National Republicans, they countered by branding Jackson a "Tyrant" and a "Usurper," who had used his presidential powers to trample the Constitution and the Bill of Rights. And, to add to the grotesque picture, they pointed to his running mate, Martin Van Buren, and asked the electorate if that nomination did not frighten them half to death.

The verdict of the American people in the fall election was a curious amalgam of victory and defeat (of sorts) for the President. The victory can readily be seen in the electoral college wherein Jackson received 219 votes; Clay, 49; and Wirt, 7. Wirt took Vermont, while Clay won Massachusetts, Rhode Island, Connecticut, Delaware, Kentucky, and a majority of the Maryland vote. Jackson garnered all the rest except South Carolina, which gave its 11 votes to John Floyd of Virginia. But the "defeat" (admittedly the word is a trifle strong) can be read out of the popular returns. Jackson received 687,502 votes, and his opponents had a combined total of 566,297. The majority for the President—which on its face looks stupendous—actually proves how damaging the Bank issue had been to his reelection. Although the total number of popular votes cast in this election increased over that of the previous election by more than ninety thousand votes, Jackson's popular majority declined from what he had received four years earlier by more than fourteen thousand votes. He is the only president in American history whose re-election to a second term registered a decline in the percentage of popular votes. Thus, despite Jackson's vaunted popularity and personal appeal, many Americans turned away from him after his first term in office!

Although it is very difficult to analyze returns of this 1832 election because they are incomplete—for example, the Missouri vote for Clay is nonexistent; the Tennessee vote for Clay is too small to be accurate; and the votes for Clay in many states are merged with Wirt's, making it impossible to arrive at a true perspective of their relative performances— nevertheless one thing seems clear: Andrew Jackson was badly hurt by the Bank issue in this election. Taken by itself, the issue could defeat a candidate, just as Clay and Biddle presumed. What spared Jackson was the overpowering strength of the Democratic party capitalizing on the President's strong personal appeal. The election was won, therefore, not by the Bank issue, but by the parades, illuminations, barbecues, "transparencies," hickory pole raisings, songs, processions, rallies, dinners, and all the other techniques the Democrats had perfected to delight the public and capture their vote. As in the election of 1828, party organization was again a major factor in the victory. "Have you an organization in your state?" asked Amos Kendall of a Connecticut politician. "Whether you have or have not . . . send me a list of names of Jackson men good and true in every township in the state . . . to whom our friends may send political information. I beg you to do this *instantly*." Then, during the campaign, the well-tooled organization functioned perfectly, particularly in taking the people's mind off the seriousness of the Bank War. For example, a torchlight procession that stretched a mile long was held in New York City. The banners were all transparencies because of the darkness: on some, were written the names of the Democratic organizations sponsoring the march; on others, stinging anathemas against "Nick" Biddle. After these came

huge portraits of the General, either on foot or on horseback, followed by those of Washington and Jefferson with
democratic mottoes surrounding the pictures mingled with
such emblems as eagles and other Roman insignias.
Periodically, the procession halted in front of the homes of
well-known Jackson men, and the crowd rendered a series
of lusty cheers and huzzahs. Then, to show how democratic
they really were, the marchers stopped at the houses of
National Republicans and razzed them with three, six, or
nine groans, depending on their rank. One contemporary
rightfully stated that these parades, as well as the other
folderol, were "episodes of a wondrous epic which will
bequeath a lasting memory to posterity, that of the coming
of democracy."

Indeed, that was one of the most important aspects of
Jacksonian Democracy: encouraging the electorate, through
party activity, to an increased participation in the political
life of the nation. And Andrew Jackson was the symbol of
that encouragement. Thus, the President, who risked his
election, his popularity and in a sense his position in history
on an issue that politically could do him nothing but harm,
was vindicated by a party apparatus that mobilized the
Democracy and rewarded him with a second term in office.

By contrast, the organization of the National Republican
party was exceedingly weak, a fact easily demonstrated by
the inability of the party to turn out Clay votes as such. In
many states—more than half a dozen at least—the Clay vote
was merged with the Anti-Masonic results and simply
reported as the "anti-Jackson" vote, as though the election
was a contest between "Jackson" and "anti-Jackson." Surely
this is no way to win an election.

In the vice-presidential contest, Van Buren easily won, with
189 votes to Sergeant's 49 and Ellmaker's 7. South Carolina
gave its 11 votes to Henry Lee of Massachusetts, while
Pennsylvania awarded 30 to its favorite son, William Wilkins.

If nothing else, the election convinced Jackson that the Bank's "absolute control over the currency . . . control over property . . . control over the people" must end immediately, that he had to go beyond a simple veto and remove the deposits of the government from Biddle's vaults, thus terminating the financial association of the government with the institution even before the old charter expired in 1836. If he did not, the Bank might successfully utilize its three remaining years to upset the verdict of the election. And that frightening possibility had to be prevented at all costs. But the President was kept from proceeding at once against the Bank by the problems of the nullification and tariff controversies during the winter of 1832–1833. Then, with the coming of spring and the working out of a tariff compromise, Jackson was finally free to withdraw the deposits.

With the decision to remove the government's money, the Bank War entered its second and more confused and complicated phase. Part of the complication was Jackson's failure to provide a banking substitute for the B.U.S. to help regulate currency and credit. Another, and more serious part, was the widespread opposition among Democrats to two things: first the removal of the deposits, and second Jackson's attempt to substitute hard money for paper money. Indeed, it was a terrible agony that the President now put Democrats through, one that severely tested their loyalty, trust, and confidence in him.[8]

Take the removal first. Although the President had the full support of Taney, Kendall, Blair, and others for his radical proposal, numerous Democrats, including the Secretary of the Treasury, Louis McLane, opposed it, and predicted a dangerous depreciation of the currency if Jackson proceeded. Since removal could not be accomplished except by the action of the Treasury Secretary, this meant that McLane had to be replaced—and this was eventually done by kicking him upstairs. William Rives, the minister to

France, was anxious to return to the United States, and the then Secretary of State, Edward Livingston, was desirous of replacing Rives; so a simple shuffle was arranged, by which Livingston moved to France, and McLane moved into the State Department. Then, at the suggestion of McLane, William J. Duane of Pennsylvania was brought in as the new Secretary of the Treasury.

Meanwhile, Jackson sent Amos Kendall on a tour of the major cities to find bankers willing to accept the government deposits, but who also qualified as Democratic partisans. With so much cash to distribute, Jackson demanded reasonable assurances that a receiving bank could pass political muster. To check the politics of possible banks, as well as to initiate preliminary negotiations, Kendall journeyed from Baltimore to Boston in the summer of 1833. His pockets figuratively stuffed with money he found any number of willing bankers who professed their undying belief in Jacksonianism. However, when the time came for the money to start flowing into these new coffers, the Secretary of the Treasury, William Duane, stoutly refused to authorize the transfer. He had reservations about the safety of the state banks and feared that the removal might shake the public credit, to say nothing of public confidence. A man of conviction, he defied the President—whereupon Jackson dismissed him. Another quick shuffle was arranged, and Roger B. Taney, the tall, angular-faced, cigar-smoking radical from Maryland, was transferred from his position as Attorney-General to the office of Secretary of the Treasury; and Benjamin F. Butler of New York, one of Van Buren's closest confidents and former law partner, was given the post of Attorney-General.

On September 25, 1833, an order went out that, commencing October 1, all future government deposits would be placed in selected state banks—called "pet banks" by the opposition newspapers. For operating expenses, the government would draw on its remaining funds from the B.U.S. until they were exhausted. In this way an easy transition from national banking to deposit banking would result—or so it was hoped.

No such luck. The action of deposit removal stirred the Bank to one last-ditch fight. "Czar Nicholas," as the Democrats dubbed Biddle, snarled his defiance. "This worthy President thinks that because he has scalped Indians and imprisoned Judges, he is to have his way with the Bank. He is mistaken." And he demonstrated what he meant by unleashing the full power of his colossus. He ordered a general curtailment of loans throughout the branches of the B.U.S., an order of such sudden constriction that the country was thrown into an economic panic reminiscent of 1819. But Biddle was fighting for the life of his Bank. It was his duty, he felt, to strike back any way he knew how. Perhaps, if he brought enough pressure to bear on the President, he could force him to restore the deposits—then, maybe with luck and if the panic persisted, Jackson would be compelled to continue the charter. Although the "pets" were now receiving the government moneys, they were unable to satisfy the many demands placed upon them by the squeeze; and soon they, too, capitulated to Biddle's policy and curtailed loans. As the panic spread across the country, the Democrats, fearing a general financial collapse, worried about a party revolt. But their fears could not sway Jackson from his determination to kill the Bank. "I have it chained," he raged, *the monster must perish.*"⁹

The second agony Jackson inflicted upon Democrats—especially among rising capitalists in search of competitive gain—was his hard money scheme. A bill, prepared by Taney, and submitted to the House of Representatives in 1834 by James K. Polk, of Tennessee, required deposit banks to cease issuing or receiving bank notes under five dollars. Later, in easy stages, they were to prohibit notes under ten dollars, then under twenty dollars. In the Senate, the proposal drew powerful and articulate support from Thomas Hart Benton who, like Jackson, insisted on gold and silver as the only circulating medium of exchange. Both men believed that specie was honest coin to be paid honest men for honest work. An elimination of paper money, said Benton in a Senate speech that required two days for delivery, would restore the country to the virtues of simple yeomen, and yet it would not interfere with businessmen pursuing their legitimate economic interests. Under a paper system, "Old Bullion" continued, the country was saddled with boom or bust financial cycles. When the economy boomed, creditors were rewarded with large returns on their investments; but when a bust resulted, credit disappeared, and the working classes had no money to pay their debts, thus losing their property. No matter what condition existed, therefore, it seemed that the rich got richer and the poor poorer—just as the President had declared in his veto.

Although the hard money bill elicited Jackson's entire approval—he even declared that "Mr. Taney, Benton, and Polk deserve not only golden medals but the gratitude of their country,"—the measure terrified segments of the Democracy, especially big city entrepreneurs and financiers, who falsely assumed the President would eventually go higher than $20 with his hard money proposal. And how could they possibly function without adequate paper, they asked. Some Tammany Hall politicians even thought of creating a "ten million dollar monster" in New York City to fill the void, make a quick killing, and capture the financial leadership of the country.

Then, in January, 1834, Jackson slashed again at Biddle's dying corporation by ordering an end to the Bank's operation of paying pensions to Revolutionary War veterans; yet, when the Secretary of War, Lewis Cass, instructed Biddle to relinquish the funds and books to special agents, "Czar

Nick" scornfully rejected the order. Again he repeated his foolish remark that the President of the United States was not going to have his way with the Bank.

It was at this point during the so-called Panic Session of Congress that the War became extraordinarily disordered and complicated. What had begun as a straightforward power struggle sank into a morass of economic and political confusion, with some Democrats advocating a new national bank and others opposing it, some favoring hard money and others violently hostile to it. (And, of course, there were some Democrats who ignored all the tumult and shouting about money and banks and simply continued to "huzzah" for Old Hickory.) During the first stage of the War many Democrats had deep reservations about the veto of the charter—which was bad enough—but now the removal of the deposits and the hard money scheme, they felt, were acts of sheer lunacy certain to cripple the country. Some of them deserted the party; others remonstrated with a President gone deaf to their entreaties; and still others plotted to create a new national bank.

One of the latter group was the Attorney-General of the United States, Benjamin F. Butler, who kept in almost daily contact with the chief banker of the Albany Regency, Thomas W. Olcott. Like Butler, Olcott held a high position in Van Buren's political machine, and as the cashier of the Mechanics and Farmers Bank in Albany he was something of a financial adviser to the Regency cronies. Writing to Olcott, Butler asked him to suggest the "best scheme" for a "new Bank to be located in Washington, with a capital not exceeding 10 millions." He said he had "no doubt as to the constitutional power of the Congress to create a Bank with such a capital & other arrangements, as might be required by the *real necessities* of the government." Coming from Jackson's Attorney-General, this was a most remarkable admission. But Butler was not alone in his views. Many

Democrats in Congress, and within the administration, were impatiently anticipating the demise of Biddle's corporation now that the deposits had been removed in order to replace it with still another national bank. And some of these men were the loudest decriers of privilege and monopoly. There were "more than thirty members" of the House of Representatives alone, reported Congressman Robert T. Lytle of Ohio, "who voted with us, & will throughout against restoring the Deposits and also against the present Bank who are *decidedly* in favour of a *New one*." "Our friends," recalled Congressman Richard M. Johnson of Kentucky, "were divided during the Panic Session upon *the Bank & a Bank*." Although several conferences were held to determine a course of action, the Democrats could not agree over the form and functions of the new bank. One group wanted a reproduction of Biddle's company, but on a smaller scale and under better government supervision. Like Butler, they envisioned a ten million dollar institution that would be located in Washington. A second group advocated "a people's Bank," in which the stock would belong to the several states and the United States. Still a third group argued for a bank in the District of Columbia that would merely serve as a "fiscal agent" for the government with no power to grant loans. Most of these Democrats did not believe that Jackson's "experiment" with deposit banking would work. Furthermore, they were apprehensive over its popular reception. "If State Banks will do," Johnson explained to Francis P. Blair, "I am better satisfied than with any National Bank—but if time should not convince the people of this & we all see it I feel confident that you & myself will concur in the course."[10]

While his friends and enemies plotted behind his back or zigzagged in the front or to the sides of him, the President maintained his steady course. All during the winter and early spring of 1834, he was badgered to give back the deposits and the charter. Delegations of businessmen from Philadelphia, Baltimore, Boston, and other leading cities caught in Biddle's squeeze pleaded with Jackson to save them from bankruptcy. They told him they were insolvent. "Insolvent do you say?" bawled Jackson as he staged a volcanic rage for their benefit. "What do you come to me for, then? Go to Nicholas Biddle. We have no money here, gentlemen. Biddle has all the money. He has millions of specie in his vaults, at this moment, lying idle, and yet you come to *me* to save you from breaking. I tell you, gentlemen, it's all politics." Indeed. And the worst kind, too; for as these disturbed and anxious businessmen filed out of the room, Jackson settled back in a chair, reached for his pipe and started laughing. "They thought I was mad," he chuckled. The old scamp. He was up to his tricks again, carrying on, flying into a rage when all along "his self-command was . . . perfect," "his passions completely under control."

There were other delegations and other phony performances, with Jackson insisting he would never recharter the Bank. The panic had been started by Biddle, he said, and if the government capitulated, it would mean that the United States would be forever subject to the dictates of the monster Bank. It was just a matter of hanging on a little while longer. He had the beast chained, he said. Soon it would be dead.

But some Democrats did not wish to wait, and together with National Republicans, nullifiers, tariff men, States' righters, Bank men, and other dissidents, they formed a new party called the Whig party to designate their opposition to concentrated power in the hands of the chief executive. The name may have been used first by James Watson Webb, editor of the New York *Courier and Enquirer,* but it received official sanction when Clay gave it his blessing in a speech delivered in the Senate on April 14, 1834. So many factions converged into this new party that about their only common agreement was their objection to the vigorous use of presidential powers by Andrew Jackson. They rejected his concept of a strong, dynamic, and aggressive chief executive. They condemned his use of presidential authority for economic and social, as well as political and diplomatic purposes. Their newspapers crowned him "King Andrew I," a despot of the Old World come to extinguish the freedoms of the New. They predicted frightful consequences to the country if he should succeed in his financial folly.

To make matters worse, there were persistent rumors in Congress that a New York conspiracy was being hatched to kill the Biddle corporation and replace it with *"a Bank to be located at N. Y."* The rumor jolted Democrats. Several members of the Pennsylvania delegation, when they heard the report, verged on open revolt against the President; so, too, Southerners, who were appalled by the arrogance of New Yorkers. "Other desertions are talked of," reported one man close to the scene, and for a time it looked as though Jackson's majority in the House of Representatives was about to break up. Actually, there was no conspiracy. Although members of the Regency were indeed considering the possibility of establishing another bank (most men on the make had something going these days) they had too much sense to locate it in New York, for they recognized that such a scheme would annihilate Van Buren's presidential prospects for 1836. Attorney-General Butler told Thomas Olcott that if the Treasury could manage its affairs without a bank, then "I hold we have no power to make one." But, if after "a *fair* trial" it should be seen that a national bank was essential, then "I have not the slight[t]est doubt of the power to make one."

Responses from other Democratic politicians also urged a new bank. Panic conditions among the merchants, said one, "is truly appalling," and "there is great danger of a total prostration of the Bank credit of the country." Yet these reports of financial stringency were highly exaggerated, and some were purposely distorted. The Bank question had become such a political football that leaders overreacted to each new report about economic distress or the kind of financial system that would eventually replace the B. U. S.

To extinguish the rumors about New York's schemes, Van Buren prevailed upon Silas Wright, his ablest lieutenant in Congress, to deliver a vigorous speech in the Senate denying a Regency takeover. Said Wright to a packed chamber: "I go against this bank and against any and every bank to be incorporated by Congress, whether to be located at Philadelphia, or New York, or any where else within the twenty-four independent States which compose this Confederacy. . . ." Such a clear and forthright statement from New York politicians produced a sobering effect on Congressmen. Coupled with Jackson's renewed declaration of unbending opposition to the Bank—any national bank— it eased the tension among Democrats and dispelled their fears about a New York or Wall Street conspiracy.[11]

Not that Democrats ceased their scheming; but unfortunately they worked at cross-purposes and did not know how to proceed without encountering Jackson's fierce opposition to paper money or to any proposal that could lay him open to the charge of destroying the B.U.S. just to replace it with a replica colored to his own political persuasion. Consequently, the Tammany men were warned away from their ten million dollar shenanigan, while Butler calmed the fears of New York financiers about Jackson's hard money ideas. "I assure you that he entertains no such Utopian dreams" as reported in the newspapers, said Butler, and if "there are any persons who wish to go further it is enough to say that their number is not only very small, but that there is not the slightest possibility of carrying any such project into effect."

Meanwhile, the Whigs in Congress, led by Henry Clay,

continued to pommel the administration with charges of employing the issue to transform the Republic into a dictatorship—of arrogating to the presidential office powers not warranted under the Constitution. In the Senate, Clay, Webster, and Calhoun formed a powerful triumvirate, and they pilloried Democrats for blindly following the "Tyrant" in the White House. One incident helped their cause. At the beginning of the session, Clay inveigled the Senate into asking the President for the document he read before the Cabinet about the removal of the deposits. Quite properly, Jackson refused the request as an invasion of his executive authority, and Clay seized on the refusal as an excuse to introduce two resolutions in the Senate: the first, censuring Jackson for the removal as a misuse of presidential power, and the second declaring the reasons offered by Taney for the removal as "unsatisfactory and insufficient." On February 5, 1834, the second resolution passed by a vote of 28 to 18; and on March 28, the censure resolution passed by a vote of 26 to 20. To add insult to injury, the Senate also refused to confirm the nomination of Roger Taney as Secretary of the Treasury.

The censure hurt Jackson deeply. Immediately he fired a "Protest" at the Senate, claiming the action utterly incompatible with the "spirit of the Constitution and with the plainest dictates of humanity and justice." "So glaring were the abuses and corruptions of the bank," he wrote, ". . . so palpable its design by its money and power to control the Government and change its character, that I deemed it the imperative duty of the Executive authority . . . to check and lessen its ability to do mischief. . . . The resolution of the Senate . . . presupposes a right in that body to interfere with this exercise of Executive power. If the principle be once admitted, it is not difficult to perceive where it will end. If by a mere denunciation like this resolution the President should ever be induced to act . . . contrary to the honest convictions of his own mind in compliance with the wishes of the Senate, the constitutional independence of the executive department would be ... effectively destroyed and its power ... transferred to the Senate...."

Shrewdly, Jackson closed this "Protest" by reaching beyond the Senate and addressing himself directly to the American people. All he had tried to do, he asserted, was "return to the people unimpaired the sacred trust they have confided to my charge; to heal the wounds of the Constitution and preserve it from further violation; to persuade my countrymen, as far as I may, that it is not in a splendid government supported by powerful monopolies and aristocratical establishments that they will find happiness or their liberties protection, but in a plain system, void of pomp, protecting all and granting favors to none, dispensing its blessings, like the dews of Heaven, unseen and unfelt save in the freshness and beauty they contribute to produce. It is such a government that the genius of our people requires; such an one only under which our States may remain for ages to come united, prosperous and free. If the Almighty Being who has hitherto sustained and protected me will but vouchsafe to make my feeble powers instrumental to such a result, I shall anticipate with pleasure the place to be assigned me in the history of my country, and die contented with the belief that I have contributed in some small degree to increase the value and prolong the duration of American liberty."

This final paragraph is persuasive evidence of Jackson's remarkable gifts as a politician and leader of his people. Also, it is as close to a description of one aspect of Jacksonian Democracy as one is likely to find by the man whose name is forever linked to the magnificent equalitarian surge that occurred in the middle period of American history. And rightly, too, is his name so linked; for Jacksonian Democracy, though diverse, complex and many-sided, was fundamentally a political movement that did two things: it preached the equality of men in their relations with the federal government; and it appealed through a variety of new techniques for increased party participation by a mass electorate. In both instances Jackson played an essential role. Indeed, the beginning of the modern party system in America is inconceivable without the name, image, and reputation of Andrew Jackson. In short, he is central to an understanding and appreciation of the entire age—as the Bank War clearly demonstrates.

For, despite the censure by the Senate, despite repeated efforts to harass the President and force recharter, despite Democratic gamesmanship, Andrew Jackson would not yield and restore the privileges or deposits to the B.U.S. Skillfully using his executive powers, he rallied his party. Though plans for a Bank substitute were still bruited about, the President calmly insisted he would veto any agency or system that concentrated the money power under private or public control, at least until his own "Experiment" with deposit banking had had a "fair trial." In addition, he patiently waited for Biddle, giving him all the time necessary to convince the country by his rashness that the Bank was truly a dangerous monopoly. Also, Jackson dispelled party fears about a northern conspiracy to fill the pockets of certain commercial cities. He was so successful in giving these assurances that by March, 1834, Butler was telling his New York cronies that another national bank was absolutely out of the question.

In asserting party leadership, Jackson was gracious and gentle at one moment, sharp and abrasive the next. With Southerners, he was generally soothing and pleasant. But with the Pennsylvania delegation, which had led the movement toward desertion from the party, he was merciless. "I am told," said a Pennsylvania Whig, "that he absolutely rode with whip and spur over our delegation who were so overwhelmed that they had nothing to say for themselves." Thoroughly chastened by the lashing, the Pennsylvania Congressmen reformed their ragged lines behind the President, moving faster when their formerly pro-Bank governor, George Wolf, publicly damned Biddle for bringing "indiscriminate ruin" upon the community. This public rebuke by the Pennsylvania governor was the turning point in the long war, and after April, 1834, Biddle's institution was forever doomed.[12]

In closing ranks under presidential leadership the Democrats now proved what stalwart party men they really were. Brushing aside minor disagreements and working closely with the White House, they jammed a series of resolutions through the House of Representatives on April 4, 1834, that renewed the Hero's spirit. The first of these disapproved recharter; the second opposed restoring the deposits to the National Bank; the third recommended the continued use of state banks for government moneys; and the fourth called for an investigation of the B.U.S. to learn to what extent the panic had been deliberately instigated by Biddle. A committee sent out under the fourth resolution to investigate "Czar Nick" found him as defiant as ever. He refused permission to examine his books and he refused to testify before the committee. Although Congress did not hold him in contempt, as it could have, his action, said one committeeman, proved "to the people never again to give themselves such a master."

The panic, which was not as severe as some businessmen and politicians pretended, soon passed, since it was artificial in its origins. Biddle, now criticized even by his staunchest supporters, was forced to loosen his financial grip, and, once freed from this restriction, the country experienced a new and unfettered entrepreneurial thrust of heroic strength and dimension. Unfortunately for Jackson, his hard money policy went by the board—for state banks, backed by government money, flooded the country with their paper. Still he continued to hope for the best. He urged his new Secretary of the Treasury, Levi Woodbury, to circulate as much gold coin as possible "to convince the people how ideal & falacious has been all the noise made by the bank men of the necessity of a national Bank to regulate the currency and being necessary as a fiscal agent of the government." But things did not go as he planned, and in time there were nearly a thousand different kinds of paper in circulation, some of it from places with such blood-curdling names as "Glory Bank." Discouraged, Jackson wistfully conjectured about creating a Bank of Deposit and Exchange in Washington, one that would report annually to Congress "its whole proceedings, with the name of all its debtors &c &c." It would also have power to issue bills over $20. He thought such a project might be a good example for the states as well as prove beneficial to the "safety of our currency . . . check the paper system & gambling menace that pervades our land & must if not checked ruin our country & our liberty." Yet, with the swift upward movement of the economy, completing the establishment of the industrial revolution in America, this currency and credit inflation constituted a tremendous boon to the country. Although sincere in his desire to reform the currency, the President was naive about the advantages of hard money and so, by his actions, unintentionally made available the mightiest ingredients of industrial growth and development.

Other benefits followed. The congressional elections in the fall of 1834 witnessed a remarkable swelling of popular support for the Democratic party. Not only did the Democrats increase their strength over the Whigs in the House, but they became the majority party in the Senate. It was a personal triumph for Jackson. Not much later, the Senate expunged the censure from the record by drawing large black lines around the offending words and writing across them, "Expunged by order of the Senate, this 16th day of January, 1837."[13]

Now, except for the business of winding up its affairs, the Second Bank of the United States was dead. The question then arises about the wisdom of killing it—whether the country was harmed or benefited by Jackson's action. From the hindsight of over a century the answer seems quite simple: in point of fact, the Second National Bank should never have been chartered in the first place—at least not without adequate government control and regulation. Lacking these, it concentrated too much power in private hands.

But give the devil his due. The B.U.S. served as an excellent central banking system and regulated credit operations efficiently and judiciously. Even so, it retained an enormous capacity for harm, a capacity repeatedly demonstrated during the life of the charter. However much it was goaded into battle by President Jackson, the fact remains that the Bank could and did retaliate by disrupting business and threatening the government. This was intolerable. Biddle's refusal to restore the veterans' pension funds to the Secretary of War, and his refusal to open his books to Congress or testify before its committee may be minor incidents, but they indicate something desperately wrong in the relationship between the government and its chartered agent, the Second Bank of the United States.

Although the federal government owned stock in the B.U.S. it did not and could not control the Bank's operation. What the nation needed, therefore, was a measure of *effective* government regulation of the entire system if a national banking institution was to serve the country as an honest broker among all classes of people. Anything less was bound to create inequities. However, in the 1830's, because of the prevailing philosophy of limited government, such regulation was probably impossible. Jackson, therefore, did the next best thing: he destroyed the Bank, root and branch.

The war to obliterate the B.U.S. was no morality play, no simple struggle between Jackson and the democracy on the one hand and Biddle and the moneyed aristocracy on the other. What the President did in his own strange and uncommon way was assert and uphold the supremacy of the federal government even as he struck down one of its chartered agents. Using the vastly strengthened presidential power through the veto, citing constitutional, political, social, and economic reasons for his action, he fought an institution that he believed had become more powerful than the central government. Then, exercising his party leadership in Congress, he finished the fight and eliminated the B.U.S. altogether. The Bank had the "money and power to control the Government and change its character," he said. This he would not tolerate. In the course of the struggle he rose above politics. The issue was not one that naturally found favor with the people or politicians because it alarmed them by threatening their security. However, he went to them and asked for their support, and when they gave him an equivocal answer he pressed ahead anyway. Jackson did not initiate the Bank War to seek or increase his popularity. He did what he believed was his sworn duty as chief executive: to terminate the ability of the Bank to "control" the American people and their government. Said he in summing up the War: "The Monster was likely to destroy our republican institutions, and would have entirely subverted them if it had not been arrested in its course."[14]

CHAPTER IX

"I Do Precisely What I Think Just and Right"

As LONG AS HE RESIDED IN THE WHITE HOUSE, PRESIDENT Jackson liked to believe that he represented the people against aristocracy and privilege, and that no group, class, or agency was entitled to any special political or economic advantage. Indeed, at the heart of Jacksonian Democracy was the commitment to the principle that all men were equal in their relation to the government. Because the rhetoric of Jacksonian newspapers pulsed with the doctrine of equality, and because the Democratic party organization worked fervently to broaden its political base among the great numbers of the American electorate, many different classes of people instinctively gravitated to the Old Hero. Mechanics, laborers, farmers, intellectuals, businessmen, and professional men, all looked to him for leadership, looked to him to keep the government from promoting any one group against the others. They did not expect him to swing the government behind the masses in their struggle for advancement; but they did expect him to maintain the government as an honest referee in the normal economic and political functions that operated within society.

It is not surprising, therefore, that during this Age of Jackson, many men, inspired by presidential leadership, sought equality for themselves and others in a wide assortment of reform societies. Frequently organizing into local and national groups, they agitated for expanded public education; the abolishment of imprisonment for debt; women's rights; care of the poor; world peace; temperance; improvement of prisons and insane asylums; and the abolition of slavery. They experimented with communitarian doctrines; consorted with philosophers, who preached the perfectibility of man and society; took up a variety of fads, from phrenology to vegetarianism; and read the new and fascinating books from the literary stalls of New York and New England. In short, the Age of Jackson was an exciting time of men searching and experimenting and changing. It was a romantic age of hope and promise.

Jackson's role in all this was more symbolic than anything else. His life and accomplishments typified the American struggle for improvement—and its highest degree of achievement. Perhaps that was one reason why the people loved him so much—this fierce, choleric old man who rumbled and growled and made a great deal of noise as he went about the business of the presidency, but who somehow got things done to the complete satisfaction of the electorate. Their affection for him exceeded that of any previous president, including Washington and Jefferson. They showed it, too, by regularly sending him gifts, canes, hats, pipes with long stems, an elegant phaeton made from the wood of the old frigate *Constitution;* and one person sent him an enormous cheese four feet in diameter and weighing 1,400 pounds. Those who visited him found not the terror they had heard about but a courtly, dignified man who looked and acted like a President and treated everyone with deference and courtesy.

Because it was so easy to gain access to the President, and because Jackson was such a controversial figure, he became a target for maniacs. The first assault on his person occurred on May 6, 1833, when the presidential party left the capital by steamboat for Fredericksburg, Virginia. At Alexandria, a former naval lieutenant, Robert B. Randolph, who had been dismissed from the service for attempted theft of funds belonging to John B. Timberlake (of all people), came aboard and made his way to Jackson's cabin, where he found the President seated at a table reading a newspaper. Jackson looked up. "Excuse my rising, sir," said the President, who did not know his visitor. "I have a pain in my side which makes it distressing for me to rise." Randolph started to remove his glove. "Never mind your glove, sir," said Jackson, extending his hand. And with that, Randolph slammed his fist into Jackson's face. The captain of the ship, who happened to be standing close by, rushed forward and grappled with the assailant, whereupon a violent scuffle ensued. Finally some of Randolph's friends appeared, pulled him off, and hurried him away from the ship before any of the other passengers knew what had transpired. Jackson, his face slightly bloody, muttered that had he known Randolph's purpose he would have defended himself. "No villain," he said, "has ever escaped me before; and he would not, had it not been for my confined situation."

A more frightening incident—the first attempt to assassinate a President—took place in January, 1835. Jackson had gone to the House of Representatives to attend the funeral services of the late Congressman, Warren R. Davis, and as he filed past the casket in the rotunda of the Capitol, a man named Richard Lawrence, who was standing about six feet from the President, drew a pistol and fired it point blank at Jackson. The report thundered in the rotunda. Before anyone could recover from the shock, the assassin pulled out a second pistol and again fired at the President. Fortunately, only the caps of the pistols exploded and Jackson was uninjured; but no sooner did the President come awake to his danger than he lunged at the assailant, his cane raised to strike. However, a young army officer reached Lawrence first, subduing him before Jackson could thrash him with his cane. Later, Lawrence claimed he was the heir to the British crown and that Jackson somehow or other stood in the way of his coronation. With that distressing announcement, the unhappy man was carted off to an insane asylum.

Despite the violence that accompanied so much of Jackson's career, he lived a very quiet life at home. In the White House, Mrs. Emily Donelson preserved the air of calm and domesticity that the President enjoyed. Frequently, there were guests of course, but they were entertained on a modest scale. Actually, people dropped in whenever they chose: sometimes before breakfast and many times late at night. At formal state receptions the President greeted all as they arrived, shaking hands with the men and bowing to the ladies. Ices, wine, and lemonade were served. Supper generally started about eleven o'clock in the large dining room of the White House with the food and drink placed on a counter or table to one side of the room. At small dinner parties, guests arrived about four in the afternoon and were treated with choice wines. Jackson ate little, usually a bit of bread, some milk and vegetables. After dinner, he accompanied his guests on a tour of the grounds, showing off his shrubs and flowers and the magnolia tree he had planted.

Jackson spent exceedingly long hours working as President. Any number of stories relate how he was repeatedly found at his desk, plugging away at his duties, reading reports, preparing others, going through his papers with an intensity that evinced a sense of tremendous responsibility. He was attentive to detail, which surprised some men who thought he was too excitable to take the time with such matters. Because sleep frequently eluded him, he stayed up half the night, sitting at his desk studying documents or writing letters. During the later years of his administration, he was often quite ill from the infirmities of age, old wounds, and weak lungs; and so he worked lying on a couch and took most of his meals in bed. When the weather in Washington grew hot and muggy, the President retired to his favorite summer retreat at the Rip Raps on Old Point Comfort at the mouth of the Chesapeake River. In winter evenings, however, he liked nothing better than to lounge in the large parlor of the White House wearing a long, loose coat and smoking his reed pipe. Nearby, four or five ladies might be found sewing, such as Mrs. Donelson, Mrs. Andrew Jackson, Jr., Mrs. Edward Livingston, and perhaps one or two others. Several children, from two years of age to seven, played over in one corner of the room, and if they made too much noise, Mrs. Donelson would hush them with a signal. A cabinet officer might be there, such as Edward Livingston, the Secretary of State, reading to Jackson in a low tone, possibly a dispatch from the French minister for foreign affairs, with the President interrupting to comment whenever the occasion warranted.[1]

Such was the informality by which government business was conducted in this sylvan age. But, however informal his approach to his duties, Jackson carefully scrutinized the myriad details essential to the proper performance of his office. Take diplomacy, for example. Jackson administered foreign affairs somewhat like he did domestic politics. He applied a combination of aggressiveness in searching out the problems that needed resolution, patience in working out details, tact, a threat of force when patience ran out, and compromise. And this combination under his direction proved eminently successful. In foreign affairs, he scored a number of notable victories that enhanced the prestige and reputation of the United States in every part of the world.

Perhaps his first striking success involved trade with the West Indies. From a very early date, Great Britain had adopted a restrictive policy with respect to U.S. trade with the West Indies. The United States retaliated by forbidding the exportation of American goods in British ships to the islands; but this reciprocal ill will hurt both nations, and during the administration of John Quincy Adams the

British government indicated a disposition to compromise. Not Adams. He mistakenly thought that British officials were under great pressure to grant concessions—so he rejected compromise. He pressed for preferential treatment of Americans entering West Indian ports, refusing, at the same time, to allow the removal of American duties on British ships engaged in the same trade. Before the mistake could be rectified and Albert Gallatin, the American minister in London, authorized to withdraw the demands, the British government broke off the negotiations and once again closed the West Indies to American ships. Gallatin tried desperately to reopen the talks, but George Canning, the British foreign minister, would not discuss it. Thus, by incredibly bad diplomacy, Adams had sacrificed a very lucrative market that Southern agricultural and Northern mercantile interests had once enjoyed.

That was where the situation stood when Jackson became President and Van Buren the Secretary of State. But political conditions had changed on both sides of the Atlantic. In Great Britain, Canning had died and was replaced by Lord Aberdeen, who took a friendlier view toward the United States. Because he believed that diplomacy was yet another area for forceful executive leadership, Jackson reopened the West Indian negotiations, informing the British as he did so that the American people had rejected the views of the Adams administration, and that the Jackson government could be expected to approach the problem with fairness and a sense of compromise. This refreshing frankness (not altogether proper since it implied ridicule towards the previous administration) was a distinct departure from normal diplomatic procedures; it forthrightly admitted the President's intention to resolve the problem quickly and amicably.

The British reacted favorably to this gesture of good will. However, negotiations dragged along at their usual snail's pace, and Jackson became increasingly irritated over the failure to produce a fast settlement. Finally, he told Van Buren that if England was unwilling to come to terms, then perhaps the United States should apply more vigorous measures. Fortunately, this proved to be unnecessary. Responding to a hint from Louis McLane, minister to Great Britain, that an act of Congress authorizing the President to grant privileges would be met with reciprocal·action, the administration made one more effort to settle the issue by obtaining the necessary legislation on May 29, 1830. Immediately, Great Britain removed her restrictions. On October 5, Jackson completed the accord by issuing a proclamation announcing that United States and British West Indian ports were open on terms of full reciprocity, without duties against ships of either nation or their cargoes.

A splendid example of Jackson's initiative in exercising forceful executive leadership to settle long-standing diplomatic problems involved the French Spoilation claims. This controversy arose from claims by American citizens against France for the destruction of property from 1803 to 1815 during the Napoleonic Wars, particularly the seizure of American ships and goods. Europeans had presented similar claims against the French and had been paid. Now Jackson felt it was America's turn. William C. Rives was dispatched to France as minister and instructed to collect what was owed. Displaying great tact, patience, and skill, he spent months talking, cajoling and urging the French to meet the just demands of his country. But the French kept putting him off with one ridiculous demand after another. Apparently they were willing to pay in good time but not before they had entertained themselves by forcing the United States to dance a diplomatic jig. That was all right with Jackson, just as long as they finally paid up and that they did so without wasting an interminable amount of time. A change in attitude finally came when King Charles X was driven into exile during the revolution of 1830 and Louis Phillippe ascended the throne. Then, on July 4, 1831, the French agreed to a treaty that stipulated payment of 25 million francs in six annual instalments. Naturally implementation depended on the approval of the treaty by the French Chamber of Deputies and the appropriation of the money. Months and years passed without the French government making a further move, and by 1833 Jackson was boiling mad. To add to the President's aggravation, Nicholas Biddle tossed in a problem to bedevil the situation still further. This occurred when the Secretary of the Treasury drew a draft on the French government through the B.U.S. for the first instalment; since the Chamber of Deputies had not appropriated the funds, the French refused payment, whereupon Biddle charged the U.S. government for protest costs, interest, and reexchange—amounting to approximately forty thousand dollars. This was outrageous. The Bank was earning interest on government deposits, for which it paid no fee, and here it was attempting to penalize the United States for the French failure to pay its legitimate debt. Although Biddle was acting within the law in charging this penalty, Jackson regarded it as impertinent and harassing. He refused to pay the money, and Biddle's action undoubtedly had something to do with the President's decision to remove the government deposits from the B.U.S. in the fall of 1833.

When Edward Livingston replaced Rives in Paris as minister, in September 1833, he immediately put pressure on the tight fisted French government for payment. Still nothing. In fact, the French Assembly actually defeated a bill to appropriate the money in the spring of 1834. His patience finally gone, Jackson concluded, "There is nothing now left for me but a recommendation of strong measures." Accordingly he ordered the Secretary of the Navy to be

ready for service, and in his message to Congress the following December, he announced that the period of treating the
French with deference and respect had ended. Peace and
friendly relations with foreign nations had always been the
policy of this country, he averred, but peace can never be
secured by surrendering American rights or permitting
solemn treaties to be abrogated. The French were guilty of
dismissing our just claims as though they did not exist, as
though this nation counted for nothing. Under the circumstances, he said, his responsibility as President demanded
that he recommend to the Congress that "a law be passed
authorizing reprisals upon French property in case provisions shall not be made for the payment of the debt at the
approaching session of the French chambers." Here was
bare-knuckles diplomacy, the kind not calculated to win
friends. Still he gave the French one final chance to pay
what they owed.

Having invited the insult, the French were appalled when
Jackson delivered it. Their minister in Washington was
recalled, and Livingston was advised in Paris to leave the
country. He complied, placing the legation in the hands of
the chargé d'affaires, Thomas Barton. Relations rapidly
deteriorated, and talk of war was bruited about by trigger-
happy patriots on both sides of the ocean.

These grave developments did not disturb the Whigs in
Congress, however, since they had already announced their
refusal to support the President in the emergency.
Predictably, Henry Clay, the chairman of the Senate Foreign
Affairs Committee, introduced a resolution in the upper
house declaring it inexpedient for the President to threaten
reprisals, no matter what the justice of the American claim.
The House of Representatives was more obliging toward the
President and passed a three-million-dollar fortifications
bill to be used as the need arose. The Senate promptly killed
it, however.

But the people supported Jackson. They thrilled to his strong arm methods, luxuriating in the comforting thought that no one was going to bully the United States while Old Hickory presided in the White House. It was another example of Jackson's uncanny instinct for the popular gesture, the action that intensified the respect and affection the American people felt toward him.

As the months passed, both nations sobered and the possibility of war—never great to begin with—sharply diminished. In April, 1835, the French Chamber of Deputies finally voted the funds to pay the debt, but added the stipulation that the money would not be remitted until Jackson tendered a satisfactory explanation of the threat contained in his message to Congress. The President refused an explanation, suspecting another French dodge, and he specifically instructed Barton in Paris to offer none. Barton was simply told to inform the French government that the Rothschilds were the American agents and that they would receive the money due the United States. If the funds were not forthcoming after a week, Barton was ordered to demand his passport, close the legation, and return home. Barton carried out his instructions, and when the French still insisted on an explanation he closed his office and left for America.

Furious over these developments, Jackson privately confided to Amos Kendall that if the French government "does not pay the debt before next congress, I will then speak to the congress in the language of a true american & of truth . . . that all intercourse ought to be closed with her until she complies with her Treaty. . . . Still I am of opinion that France will pay the money without apology or explanation—*from me, she will get neither.*"

Yet even though Jackson's general attitude remained truculent, he was already moving toward compromise by preparing a statement to be delivered in his annual message to Congress, denying any previous intention "to menace or insult" the French government. Still he insisted that the honor of his country forbade any apology for stating the truth and performing his duty. The French, anxious for an escape from this unseemly imbroglio, chose to interpret Jackson's good intention as the explanation they demanded and immediately started payment on the debt. By the spring of 1836, the President happily announced that four of the six instalments had been paid, and that friendly relations with France had been re-established.

It was a splendid victory, and the American people burst out in a prideful round of applause at this latest Jackson triumph. The timing was perfect, too. It came just as the presidential election of 1836 was about to begin. The Whigs had hoped to capitalize on the President's roughhouse diplomacy by appearing as the party of peace and moderation, but Jackson's achievement was so stunning, and the national pride it engendered so heartfelt, that the issue was completely lost to them.[2]

Jackson enjoyed other diplomatic successes. U.S. claims against Denmark were favorably settled in 1830 with an award of $650,000 and also against the Kingdom of Naples in 1832 when Jackson agreed to accept $2,119,230 to be paid in ten instalments with 4% interest. Because of the murder of an American crew in Sumatra, the President commissioned Edmund Roberts, a merchant and sea captain, as a special agent to draw up commercial treaties with Cochin China, Siam, Muscat, and Japan. Roberts failed in Cochin China because, Jackson-like, he refused the kowtow, and he died before reaching Japan; but prior to his death he signed treaties with Siam and Muscat in 1833, opening these countries to American trade on the basis of the most-favored nation. These were the first treaties between the United States and Far Eastern countries. Jackson also obtained the first treaty between the United States and

Turkey in 1831, again on the basis of most favored nation, which permitted American merchants ships to penetrate the Bosporus into the Black Sea and provided for extraterritorial juridical status for Americans residing in Turkey.

In attempting to settle the boundary line of the state of Maine with Great Britain—a dispute reaching back to the Peace Treaty of 1783 that ended the American Revolution—Jackson was less successful. Earlier, in 1827, the matter had been turned over to the King of the Netherlands for arbitration, but the evidence supporting American claims was inconclusive, and in January, 1831, the King proposed a compromise that essentially split the difference between the two countries. Great Britain expressed willingness to accept the award, but Maine protested and the Senate voted disapproval of the King's recommendation. In May, 1833, Jackson discussed the proposed settlement with his Cabinet, and during the conversation a map was brought out showing the proposed boundary line. Secretary Livingston voiced a note of caution, warning that there might be a clamor if the boundary were adopted. Jackson wheeled around. "I care nothing about clamors, sir," he said. "Mark me! I do precisely what I think just and right." However, in view of the attitude of Maine, the nationalistic sentiments of the people, and the obvious difficulty of getting the treaty through the Senate, Jackson decided it was "just and right" to reject the recommendation of the King of the Netherlands.

In one area of foreign affairs, Jackson sustained a notable failure. He itched to buy Texas from Mexico and authorized the American minister to Mexico, Joel R. Poinsett, to pay as high as five million dollars for the entire territory right up to the great desert. But, displaying singular ingratitude for Jackson's "generous offer," Mexico refused to slice away a section of its territory. Then, because Poinsett had attempted to interfere in Mexican politics in the interest of more democratic government, the legislature of Mexico voted his censure, thus forcing Jackson to replace him. As a substitute, the President sent Anthony Butler. Now Jackson had made some pretty ghastly appointments in his time but perhaps few were as spectacularly bad as Butler's. Once inside Mexico this fast-stepping expansionist engineered one shady deal after another to snatch Texas for his chief. He tried bribery, blackmail, loan-sharking—all to no avail. When he suggested that the President seize eastern Texas by force and place him "at the *head of the country*," Jackson wrote on the back of the dispatch: "A. Butler: What a scamp." Although the General mouthed pious sentiments about honesty in the negotiations he did nothing to restrain his minister who was as corrupt a negotiator as this nation ever employed in its foreign service. Butler was allowed to remain as minister for nearly seven years—perhaps in the hope that one of his frauds would work and plop Texas into the lap of the President. When it became blindingly clear that his negotiations were a failure, he was recalled, and Powhatan Ellis of Mississippi was sent to replace him. Ellis quickly sized up the situation in Mexico and made no further effort to purchase Texas.

However, the posture of the United States in the Southwest continued to improve despite the blundering diplomacy. During the last years of Jackson's administration, thousands of Americans poured into Texas. In self-defense, Mexico protected itself with a show of force but succeeded only in provoking a revolution. The fighting began in 1835 with Sam Houston, the President's protégé, in command of the Texas army. To lend support to the rebels, Jackson ordered the commanding officer of the Western Department, General E. P. Gaines, to proceed first to the Texas border—for the ostensible reason of protecting the United States from possible Indian attacks—and then to cross the Sabine River and penetrate to the Nacogdoches, fifty miles inside Texas. Meanwhile, Houston tangled with

General Santa Anna, the president of Mexico, at the Battle of San Jacinto and decisively defeated him. Immediately, Texas appealed to the United States to recognize its independence, or to annex it.

His goal so close, Jackson hesitated. He was a nationalist straight to the bone but he was also a realist, and annexation gave him pause. For one thing, it might produce a war with Mexico; for another, it might fissure the Democratic party over the question of slavery in the area. Inasmuch as 1836 was a presidential election year and his friend Van Buren was the nominee of the party, Jackson hesitated before taking any precipitous action which could split the party and impair Van Buren's prospects for election. But, more important, slavery had developed into a new and powerful issue during the 1830's, and the President feared that the Texas question, if exposed during an election, might set off a fearful row between slavery and anti-slavery forces that could imperil the Union. So he did nothing until after the election, and then, on March 3, 1837, the day before he left office, he officially recognized Texas' independence.[3]

In all of these negotiations with foreign nations Jackson tended to overshadow his Secretary of State—the last one being John Forsyth of Georgia, who succeeded Louis McLane to the office in 1834. It was characteristic of the General to dominate his Cabinet; strong presidents usually do. Apart from everything else, the members of his official family were hardly a notch above mediocrity, and the few good ones he had, like Van Buren and Taney, had been forced out of the Cabinet by the spite of their political adversaries. To repair the injury done the Red Fox and also gratify his own need to reward loyal service, the President made him the Vice-President in 1832; he planned a further reparation in 1836 when he would confer the presidency on the Magician. As for Taney, Jackson had to wait longer, but with the death of John Marshall he rewarded him at last with the nomination as Chief Justice of the United States. The Senate confirmed him in March, 1836, with John C. Calhoun—still fighting on—voting against confirmation.

In addition to Taney, Jackson chose five new associate justices of the Supreme Court: John McLean of Ohio, Henry Baldwin of Pennsylvania, James M. Wayne of Georgia, Philip P. Barbour of Virginia, and John Catron of Tennessee. With these appointments, Jackson not only remade the court in personnel (and gave it a distinctly Southern cast), but he markedly altered its approach to social, political, and economic questions. Where the Marshall court had emphasized property rights and the sanctity of contract, the Taney court upheld the right of popular majorities acting through their legislatures to regulate property rights and the privileges of corporations. In other words, the rights of the individual and the community took precedence over corporate rights. In the case of Charles River Bridge *vs.* Warren Bridge, for example, Justice Taney, speaking for the court, held that the state retained authority to regulate corporations in order to achieve community welfare; for whenever private rights and community rights conflict, he said, community rights come first. Yet, for all its anticorporate sentiments, the decision actually aided the rapid development of industry in the United States by denying monopolistic claims of older companies, whose charters conferred exclusive privileges.

Among Jacksonians like Taney, there was a tendency to regard wealthy and powerful corporations as a prime source of corruption and inequality in society. These sentiments were probably best expressed in the literature of the Locofocos, a radical movement within the Democratic party and so named because a group of them used "locofoco" matches to light candles at a meeting held on October 29, 1835, in New York City when the more conservative Democrats turned off the lights. The Locofocos represented small businessmen, shopkeepers, workers, artisans, and lawyers, who argued for free trade and the elimination of paper money, imprisonment for debt, monopolies, and the granting of any subsidies or favors to private business. In short they advocated total laissez faire, both foreign and domestic. Unfortunately, their activities drove a deep wedge into the Democratic party, splitting it into radical and conservative wings.

In the matter of debt, the Locofocos opposed imprisonment for those unable to pay, but they also opposed any law which would interfere "in any shape with any contract of debt, either to enforce or to annul it." And for those like the Locofocos, who advocated limited government, the elimination of the national debt was an urgent need—a concept Jackson totally approved. So it was with great pride that he announced in his message to Congress on December 1, 1834, that the national debt would be discharged by the first of January, leaving a balance in the Treasury of approximately $440,000. For the first, last and only time in American history, the nation did not owe a cent to anyone. The revenues from the tariff and land sales were so large and brought such tremendous annual surpluses, that it did not take any special action of the President to achieve this "highly flattering" result. He just happened to be in office when events, prosperity and the "enterprise of our population" all converged to free the nation of its debt. Of course Jackson had advocated debt reduction for decades, and his first message to Congress expressed a "fervent hope" that this feat would be one of the accomplishments of his administration. Now, thanks to the tariff and thanks especially to land sales, his great dream was realized.

The increase in the land revenues, at first so encouraging to the President, later gave him cause for alarm. Within two years after announcing the elimination of the public debt, land sales quadrupled. They jumped from approximately four million dollars in 1834 to over 20 million in 1836. Worse, speculators acquired enormous blocks of western lands, paying for them with any paper money they could lay their hands on and bribing officials and exerting political pressure whenever necessary. Soon, land speculation reached fantastic proportions (doing a land office business became a common expression because of it), and banks extended loans and increased their notes far beyond their resources in order to ride with the inflationary surge. Jackson, no friend to speculators, and totally committed to hard money, sought to check paper inflation by ordering land offices to refuse paper money under five dollars in the payment of public lands. This order was issued by the Treasury in April, 1835; the following July 11, 1836, Jackson went the whole route by declaring specie the only acceptable money for the purchase of public lands. This Specie Circular, a well-intentioned and forceful exercise of government authority, brought a sudden end to the danger of wild and uncontrolled speculation. But the full effects of the irresponsible struggle for the quick dollar could not be shielded by executive action, and within a year a full blown depression swept the nation.

For the remainder of Jackson's term in office, however, the problem of the Treasury surplus continued. The tariff, the other source of excessive revenue, could not be changed because of the 1833 compromise, which continued in operation until 1842. Finally the Congress and the administration hit upon a scheme whereby the surplus over five million dollars was to be deposited with the state governments in proportion to the number of their representatives in Congress. Best of all, as far as the states were concerned, there was a tacit understanding that the money would never be repaid. The gift or grant—called "deposits" to avoid raising constitutional problems in the President's mind—would be paid in four quarterly instalments. Though distressed by certain provisions of the measure, Jackson signed the Deposit Act in June, 1836. Three instalments were paid before the Panic of 1837 canceled the remaining gift.

The timing of this act was significant—or so it seemed to some critics. During a presidential election year, Andrew Jackson was giving money away—free, just like Santa Claus. What a great statesman, chorused the Democrats. He solved two problems at once: relieved the embarrassment of riches in the treasury, and greased the election of his successor Martin Van Buren.[4]

In many ways, the election of the Little Magician constituted a third election for Andrew Jackson. Because of loyalty and service, Van Buren was the old man's choice, and to ratify that choice the Democratic party was summoned to convene in Baltimore on May 20, 1835. By unanimous vote, Jackson's will was translated into the predicted nomination; and for the vice-presidency the delegates selected Richard M. Johnson of Kentucky, a bad choice, but again one designated by the man in the White House.

The Whigs, despite the precedence of the last election, held no convention. Instead, various states nominated several different candidates to take advantage of all forms of regional hostility to any aspect of Jacksonianism in the hope that this action might split the vote sufficiently to send the election into the House of Representatives. Their candidates included William Henry Harrison of Ohio, Hugh Lawson White of Tennessee, and Daniel Webster of Massachusetts.

But times were good, the party was well-tooled and greased, Andrew Jackson was the hero of the people, and he had decreed that Van Buren should succeed him. So be it. In a clear-cut victory, Van Buren received 170 votes as against 24 for all of his opponents. In the vice-presidential race, Johnson encountered unexpected trouble. Facing him were such Whig candidates as John Tyler of Virginia, William

Smith of Alabama, and Francis Granger of New York. None received a majority of votes, so the election, for the first and only time, went to the Senate, where Johnson was chosen on the first ballot over Francis Granger by a vote of 33 to 16.

The election of Van Buren and the obvious appeal of Jacksonian Democracy for the American people were sources of great satisfaction to the mellowing General. In his Farewell Address, issued on March 4, 1837 and written with the assistance of Chief Justice Taney, he expressed his joy at presiding over the nation at a time when special privilege could be struck a lethal blow. "But you must remember, my fellow-citizens, that eternal vigilance by the people is the price of liberty, and that you must pay the price if you wish to secure the blessing." He enumerated what he regarded as the major accomplishments of his administration, but, like the pragmatist he was, he did not loiter with history. He gestured toward the future. He spoke of the priceless value of the Union and warned against the increasing danger of sectionalism. Although the address lacked much of the fire of some of his political messages, he ended it on a note of eloquence and triumph. "My own race," he said, "is nearly run; advanced age and failing health warn me that before long I must pass beyond the reach of human events and cease to feel the vicissitudes of human affairs. I thank God that my life has been spent in a land of liberty and that He has given me a heart to love my country with the affection of a son. And filled with gratitude for your constant and unwavering kindness, I bid you a last and affectionate farewell."

Several Whig newspapers shouted their praise to the great God above that this was the last official utterance from the worst demagogue the nation had ever known. "Happily it is the last humbug which the mischievous popularity of this illiterate, violent, vain, and iron-willed soldier can impose upon a confiding and credulous people."

Most Americans, then and later, disagreed with this dys-

peptic judgment. And the crowds who turned out for the
Van Buren inauguration in 1837 proved it by the heartfelt
tribute they tendered their old, worn-out champion. Nearly
twenty thousand people attended the ceremonies, but they
seemed less interested in welcoming the new administration
than in expressing their love for the old. While the ceremonies were in progress the people stood riveted to their
places, profoundly silent. "It was the stillness and silence of
reverence and affection," wrote Senator Benton, "and there
was no room for mistake as to whom this mute and impressive homage was rendered. For once, the rising was eclipsed
by the setting sun."⁵

CHAPTER X

Retirement

THREE DAYS AFTER THE INAUGURATION, JACKSON STARTED FOR home. Taking every occasion to rest, stopping often, visiting friends along the route, he took several weeks to reach the Hermitage in Tennessee. Cheering crowds congregated wherever he appeared, and their obvious affection stirred the old man. "This surely, going out of office as I was," he wrote, "and opposed throughout my administration by the talents wealth & money power of the whole aristocracy of the United States, but nobly supported by the Democratic republicans—the people . . . was filling the summit of my gratification, and I can truly say, my ambition."

Jackson was now seventy years old, a very infirm old man who was rarely free from pain. When he arrived home, he claimed he had only ninety dollars in his pockets. All his salary and most of the income from his cotton crop were gone.

During his prolonged absence in Washington, his plantation had been managed by his son, Andrew, Jr.—and managed none too well, according to Jackson. Once home, he immediately set about putting his plantation to rights and extricating himself from a $7,000 debt. His financial situation was aggravated by a fire in 1834, in which the Hermitage and its contents were partially burned, and by the tremendous amount of money he spent to restore his home. Although Jackson tended to exaggerate his poverty—he had a large plantation and many slaves, hardly the signs of impending bankruptcy—there is no question that his son had mismanaged the plantation and endangered his solvency. Worse, his son lied to keep him from knowing how extensive were the personal debts he had incurred, among which were notes endorsing debts of other Tennessee spendthrifts. Jackson was not legally responsible for these debts and could have invoked the new bankruptcy law on behalf of his son. Instead, he raised the money to pay the debt and then, in a characteristic gesture of confidence, he purchased Halcyon Plantation on the Mississippi River and turned it over to Andrew to manage, encouraging him to begin again, hoping he had profited by his mistakes.

It was not easy to manage money by the time Jackson arrived at the Hermitage, for in the spring of 1837 an intense and pervasive economic panic spread across the nation. Although the panic in the United States was part of a world-wide depression, angry critics blamed it on the excessive expansion of credit, currency, and marketable goods that occurred during the 1830's, on the Deposit Act that encouraged the states to a renewed program of improvements expansion, especially in transportation, and on the Specie Circular. As the depression deepened during the summer and fall of 1837, banks of New York suspended specie payments, followed by banks in other states. Hundreds of businesses failed, unemployment rose, prices fell, and bread riots broke out in several large cities. Van Buren called the Congress into special session and set before it a program in which the most important feature was a proposed bill to set up an Independent Treasury by which government funds would be placed in separate repositories in Washington, New York, Philadelphia, Baltimore, Boston, and other cities to be withdrawn as needed. Banks would no longer have access to government funds. The bond between the government and private banks would be forever severed.

Jackson heartily endorsed this Divorce Bill, for it resolved the problem of government partnership with bankers in the use of public moneys. However, it was three years before the opposition of Whigs and conservative (pro-Bank, paper money) Democrats could be overcome and the Independent Treasury bill enacted into law. Those other parts of Van Buren's program requiring legislative action received even rougher treatment from Congress, however, and the President was forced to rely upon executive order to put through measures of economic and social reform. In one of the most interesting and enlightened presidential moves, one that would do credit to Jackson himself, Van Buren ordered the establishment of a ten-hour workday for all employees on federal projects.

But try as he might Van Buren could not reverse economic events or bring the depression to a halt. Short of massive assistance to the states, business, and individuals there was nothing to do but wait out the calamity and then start again. Still a victim had to be found to expiate the sufferings of the American people, and in the presidential election of 1840 Van Buren became the sacrificial offering. He was resoundingly defeated by the Whig candidate, William Henry Harrison, who supposedly killed Indians, just like Jackson. The electoral vote was 234 to 60, but the popular vote came surprisingly close: Van Buren received 1,127,781, Harrison 1,274,624.

The Old Hero sagged when he read the returns. It was another blow in a long series of recent disappointments to this tough old campaigner. But he soon brightened, for he was essentially an optimist, and he was telling his friends to rally around Van Buren in 1844 and elect him by a thumping majority. He said he lived for the day when Democrats would return to power.

That day came sooner than anticipated. Within a month after his inauguration, Harrison died of pneumonia and was succeeded by his Vice-President, John Tyler of Virginia. A States' righter of strong convictions, Tyler flatly opposed Henry Clay's program of public works, a new national bank, and a higher tariff. As a consequence, the Whigs began battling with their new President almost immediately after he took office. Grinned Jackson: "A kind and overruling providence has interfered to prolong our glorious Union."[1]

Tyler's administration provided Jackson with two personal satisfactions: one was the annexation of Texas during the final moments of Tyler's term, the other was the remission of $1,000 (plus interest) paid by the Hero for contempt of court in New Orleans in 1815. Senator Linn of Missouri introduced the legislation early in 1842, and while the Whigs fought it out of a sense of revenge, it finally passed on February 16, 1844. Jackson wanted the money, too. It served as a kind of vindication for his personal honor, and he could use it to help pay off his son's ever increasing debts. But of the $2,732 he received, he reserved $20 of it to buy an "outfit" for "Miss Emuckfau" who was with foal by Priam.

On the question of Texas annexation, Jackson was chagrined to learn that his old friend, Van Buren, had announced against it. Both Clay and Van Buren, the ostensible candidates in 1844, opposed annexation. "I am quite sick really, and have been ever since I read V.B. letter," Jackson wrote to Francis Blair. "Texas [is] the key to our future safety. . . . We cannot bear that Great Britain have a Canedy on our west as she has on the north."

Clay received the Whig nomination despite his stand, but the Democrats had "sober second thoughts" about nominating an anti-expansionist, and at their convention chose the first "dark horse," James Knox Polk of Tennessee—"Young Hickory." The following November, the Democrats defeated Clay on a platform demanding Texas and Oregon, and this heartwarming result revived some of the old sparkle in Jackson. To know that the country had been spared the presidency of Henry Clay was comfort indeed.

It was his only comfort. For the rest, he was constantly

plagued by the mind-killing indebtedness of his son. Also, there was physical agony each day. His head never stopped aching. He coughed incessantly, for one lung was consumed entirely with tuberculosis, the other badly infected. In addition, his body was beginning to swell with dropsy. Yet, somehow this gallant man kept going. As he sat in his chair entertaining visiting politicians or old friends, his head throbbing, his lungs literally killing him, he would tell those around him how much pleasure he derived in having his family close to him, seeing so many nephews and nieces, grandnephews and grandnieces each day and watching them frolic around the plantation. And there was always the comfort of walking into the garden and visiting Rachel's grave. As his life slowly ebbed away, word was brought to him that the retiring President, John Tyler, acting under a joint Congressional resolution, had invited Texas to enter the Union as a state. When he heard this joyful news Jackson smiled. "All is safe at last."[2]

Despite the old man's acute suffering (soon he could not lie down because of the pain and had to be propped up in bed), he was pestered daily by office-seekers who wanted a favor, or his name on a petition to recommend them for a job, or some other assistance. "I am dying as fast as I can," he said with some of his old spunk, "and they all know it, but they will keep swarming upon me in crowds, seeking for office—intriguing for office."

On June 2, 1845 the dropsy caused him such distress that surgical aid from Nashville was urgently summoned. The operation, performed by Dr. Esselman, gave Jackson much relief but left him prostrate and debilitated. Though drugs were freely administered, such as they were, he could not sleep. In the ensuing days there were periods of relative calm when the pain subsided, and during these moments he enjoyed talking politics, conversing about Texas and Oregon and the possibility of war with Mexico. A politician to the end, he kept up his vast correspondence, writing almost to the day of his death, writing to some loyal friend or the new President about political affairs.

On Sunday, June 8, Dr. Esselman came to visit the seventy-eight-year-old man and found him sitting in his armchair. The doctor took one look at his patient and knew he was dying. A little later Jackson was moved from his chair to his bed, but as they lifted him he fainted. When he revived, he heard the moans of the servants outside his room. He turned to his family and friends and asked them not to grieve. "My dear children, and friends, and servants, I hope and trust to meet you all in heaven, both white and black." Then he turned and just stared at his granddaughter, Rachel.

Around noon, Major Lewis arrived. "Major," said the dying man, "I am glad to see you. You had like to have been too late." A short time later, after Jackson had lapsed into a long silence, his son took his hand and whispered into his ear, "Father, how do you feel? Do you know me?"

"Know you?" he responded, "yes, I know you. I would know you all if I could see. Bring me my spectacles."

The eyeglasses were adjusted to his head. He spoke of his impending death, whereupon all present began weeping, causing Jackson great distress. "What is the matter with my dear children?" he asked. "Have I alarmed you? Oh, do not cry. Be good children, and we will all meet in heaven."

He closed his eyes and lay still for a half an hour, breathing softly. At six o'clock his head fell forward and was caught by Major Lewis. Lewis listened for a breath but heard none. Jackson died quietly—without a struggle and apparently without pain.

Two days later he was buried in a grave by the side of his beloved Rachel. Three thousand persons gathered on the lawn in front of the big house and listened to the eulogy of the Reverend Doctor Edgar. They sang several of Jackson's favorite hymns, after which they took the body to the garden and placed it in the tomb the General had prepared for it long ago.

The services concluded, the mourners turned away. But never would they forget that they had been privileged to know General Andrew Jackson.[3]

Notes

In order to save space several citations have been combined into a single reference and placed at the end of each appropriate section. In addition the locations of manuscript sources have been reduced to the following abbreviations:

CUL   Columbia University Library
LC   Library of Congress
HUL   Harvard University Library
MHS   Massachusetts Historical Society
ML   Pierpont Morgan Library, New York City
NYHS   New York Historical Society
NYPL   New York Public Library
NYSL   New York State Library, Albany
PUL   Princeton University Library
TSL   Tennessee State Library, Nashville

CHAPTER I

1. For the charges and countercharges made during the election, see the United States *Telegraph*, January 26, February 16, March 13, May 20, July 2, 23, 1828; and the *National Journal*, September 4, 1828, May 26, 28, June 2, 16, 1827. James Parton, *Life of Andrew Jackson*, III, 141; I, vii.

2. *Telegraph*, January 26, 1828; "The Jackson Family," Andrew Jackson, *Correspondence*, I, 1–2; Parton, *Jackson*, I, 46–58.

3. The controversy can be traced in Elmer Don Herd, Jr., *Andrew Jackson, South Carolinian;* and Max F. Harris, *The Andrew Jackson Birthplace Problem*. Jackson's testimony concerning his birth can be found in Jackson to James H. Witherspoon, August 11, 1824, Jackson, *Correspondence*, III, 265; and "Jackson's Will," dated June 7, 1843, Jackson, *Correspondence*, VI, 222. See also Marquis James, *The Life of Andrew Jackson*, footnote 17, pp. 791–97; and James S. Bassett, *The Life of Andrew Jackson*, p. 6.

4. Parton, *Jackson*, I, 64; John H. Eaton, *The Life of Andrew Jackson*, p. 10; Robert V. Remini, *The Election of Andrew Jackson*, p. 62; Tracy M. Kegley, "James White Stephenson: Teacher of Andrew Jackson," *Tennessee Historical Society*, VII (March, 1948), 38–51.

5. Augustus C. Buell, *History of Andrew Jackson*, I, 38. For a criticism of Buell as a historian, see Milton W. Hamilton, "Augustus C. Buell, Fraudulent Historian," *The Pennsylvania Magazine of History and Biography* (1956), LXXX, 478–92.

6. Eaton, *Jackson*, p. 10; Amos Kendall, *Life of Andrew Jackson*, I, 14; Parton, *Jackson*, I, 72.

7. Parton, *Jackson*, I, 75–76, 89, 90, 94; "Jackson's Memorandum on his Imprisonment at Camden, S. C.," Jackson, *Correspondence*, I, 2–4.

8. Kendall, *Jackson*, I, 58; Parton, *Jackson*, I, 96.

9. Kendall, *Jackson*, I, 68; Philo Goodwin, *Biography of Andrew Jackson*, p. 6; Parton, *Jackson*, I, 100, 104–5.

10. Parton, *Jackson*, I, 107–9, 105.

11. Harris, *Jackson Birthplace Problem*, p. 55; Court Records of Surry and Randolph counties, North Carolina; and Jackson's commission to practice law, Jackson, *Correspondence*, I, 406; Parton, *Jackson*, I, 113; Thomas P. Abernethy, *From Frontier to Plantation in Tennessee*, pp. 2–32, 64–90.

CHAPTER II

1. Parton, *Jackson*, I, 122–23, 161–62; Abernethy, *Frontier to Plantation*, p. 123; Jackson to Avery, August 12, 1788, Jackson, *Correspondence*, I, 5: S. G. Heiskell, *Andrew Jackson and Early Tennessee History*, pp. 105–12.

2. Parton, *Jackson*, I, 135–75; James, *Jackson*, pp. 47–61.

3. On the circumstances of Jackson's marriage, see John Overton's statement in *Niles Weekly Register*, June 9, 1827; Lewis to Thomas Cadwalader, April 1, 1827, Jackson-Lewis Papers, NYPL; Parton, *Jackson*, I, 145–54. There is a letter by Robards to John Coffee, 1803, Coffee Papers, Tennessee State Library which strongly suggests that Robards was mentally unbalanced.

4. Jackson to Hutchins, April 18, 1833, Jackson, *Correspondence*, V, 60. Jackson's repeated financial failures raise questions about his business acumen; and his money problems following Rachel's death are strong indications of her superiority in business matters. See also Harriett S. Arnow, *Flowering of the Cumberland*, pp. 48–49.

5. Parton, *Jackson*, I, 157, 158, 172; Abernethy, *From Frontier to Plantation*, p. 137; *Journal of the Proceedings of Convention . . . at Knoxville . . . 1796*, pp. 29 ff.

6. Jackson to Overton, June 9, 1795, Jackson, *Correspondence*,

I, 14; Jackson to James Jackson, August 25, 1819, *Correspondence*, II, 427. See also the Jackson-Allison land deal, *Correspondence*, I, 21; Jackson to Lewis, July 16, 1820, Jackson-Lewis Papers, NYPL; Parton, *Jackson*, I, 196.

7. Jackson to Nathaniel Macon, October 4, 1795; Jackson to Overton, February 3, 1798; Jackson to John Jackson, June 18, 1805; Jackson, *Correspondence*, I, 17, 45, 114–16; Jackson to James Jackson, August 25, 1819, *Correspondence*, II, 428.

8. Jackson to Sevier, May 8, 1797; David Campbell to Jackson, January 25, 1802; Jackson, *Correspondence*, I, 33, 60. Bassett, *Jackson*, 76; Jackson, *Correspondence*, I, 61, note 1, and 65, note citing Tennessee *Gazette*, July 27, 1803; Parton, I, 164; Jackson to Sevier, October 9, 1803, Jackson, *Correspondence*, I, 74–75.

9. Parton, *Jackson*, I, 265–306; see statements from various observers dated February, 1806 in Jackson Papers, LC.

10. Arnow, *Seedtime on the Cumberland*, p. 275, note 103; Jackson to George W. Campbell, April 28, 1804; Jackson to John Coffee, April 28, 1804; Jackson, *Correspondence*, I, 90–91. For evidence of Jackson's early political skills, see John Coffee to Jackson, April (?), 1797; Macon to Jackson, February 14, 1800; Jackson to William Dickson, September 1, 1801; Jackson, *Correspondence*, I, 28, 56, 58–59.

11. "Early Connection with Masonry," "List of Negroes For Jackson, Jackson, *Correspondence*, I, 59, 9; Arda Walker, "Andrew Jackson: Planter," *The East Tennessee Historical Society's Publications*, 1945, no. 15, pp. 30–31; "Affadavit," January 3, 1801, Jackson Papers, LC; Jackson to Coffee, January 7, February 28, 1804; Jackson to Thomas Watson, January 25, 1804; Jackson to John Hutchins, March 17, 1804; Hutchins to Jackson, April 16, May 23, 1806; Jackson, *Correspondence*, I, 80–83, 84–86, footnote 1, 141, 145–46.

12. Jackson to Hutchins, April 7, 1805; Jackson to N. Davidson, August 25, 1804; Jackson to Arbitrators, February 29, 1812; Jackson, *Correspondence*, I, 111, 109, 217–20; Rachel Jackson to Andrew Jackson (January, 1813?), Jackson Papers, LC.

13. Jackson to W. C. C. Claiborne, November 12, 1806, Jackson, *Correspondence*, I, 153; James B. Ranck, "Andrew Jackson and the Burr Conspiracy," *Tennessee Historical Magazine* (1930), pp. 17–28; Jackson to ?, September 25, 1806; Jackson, *Correspondence*, I, 149–50.

14. Walter F. McCaleb, *The Aaron Burr Conspiracy*, p. 75;

Jackson to Jefferson, November 12, 1806; Jackson to Claiborne, November 12, 1806; Jackson, *Correspondence*, I, 156, 153; Abernethy, *Burr Conspiracy*, pp. 111–12; Joe Gray Taylor, "Andrew Jackson and the Aaron Burr Conspiracy," *The West Tennessee Historical Society Papers* (1947), no. 1, p. 89; Jackson to Henry Clay, October 27, 1806; Jackson, *Correspondence*, I, 151. See also Jackson's letters to Henry Dearborn, W. C. C. Claiborne, George W. Campbell for January, 1807, Jackson Papers, LC.

CHAPTER III

1. Bassett, *Jackson*, 79; Jackson to William Blount, July 10, 1812; Jackson *Correspondence*, I, 230; Jackson to Blount, November 11, November 13, December 31, 1812; Jackson, *Correspondence*, I, 238–39, 239–41, 243, 252–53.

2. Parton, *Jackson*, I, 368, 372; Rogers W. Young, "Andrew Jackson's Movements on the Lower Natchez Trace During and After the War of 1812," *The Journal of Mississippi History* (1948), x, 90–92. Jackson's General Orders, December 13, 1812, Jackson Papers, LC; Wilkinson to Jackson, January 22, 25, 1813; John Armstrong to Jackson, February 5, 1813; Jackson to Wilkinson, March 22, 1813; Jackson, *Correspondence*, I, 273–76, 299. Jackson to Armstrong, March 13, 1813, Jackson Papers, LC.

3. Jackson to Rachel Jackson, March 15, 1813; Jackson to Benton, July 19, 1813; Benton to Jackson, July 25, 1813; Jackson, *Correspondence*, I, 296, 310–314. Parton, *Jackson*, I, 389.

4. William N. Chambers, *Old Bullion Benton*, p. 53, note; Parton, *Jackson*, I, 395; Donelson to Coffee, January 26, 1832, Donelson Papers, LC; James, *Jackson*, p. 591.

5. Blount to Jackson, September 25, 1813; Jackson to Blount, November 4, 1813; Jackson, *Correspondence*, I, 321–22, 341; Parton, *Jackson*, I, 447, 439; David Crockett, *Life of David Crockett*, p. 75; S. G. Heiskell, *Andrew Jackson and Early Tennessee History*, p. 348. Jackson to Blount, November 15, 1813; Jackson to Thomas Pinckney, December 3, 1813; Jackson, *Correspondence*, I, 348–50, 367.

6. Jackson to Thomas Pinckney, March 14, 1814, Jackson, *Correspondence*, I, 481; Jackson to Eaton, July 27, 1827, Jackson-Lewis Papers, NYPL. On Jackson's relations with the Tennessee militia, see his correspondence with the Secretary of War, John

Armstrong, in *American State Papers, Military Affairs,* III, 785–93. Parton, *Jackson,* I, 495.

7. Jackson to Pinckney, March 28, 1814; Jackson to Blount, March 31, 1814; Jackson to Rachel Jackson, April 1, 1814; Jackson, *Correspondence,* I, 488–94. Benjamin Hawkins to the Secretary of War, June 21, 1814, *American State Papers, Indian Affairs,* I, 859. R. S. Cotterill, *The Southern Indians,* p. 189.

8. Monroe to Jackson, September 27, 1814; Jackson to Armstrong, August 30, 1814; Gonzalez Manrique to Jackson, August 30, 1814; Jackson to Manrique, November 6, 1814; Jackson to Armstrong, August 5, 1814; Jackson to Monroe, November 14, 1814; Jackson, *Correspondence,* II, 60–62, 36–37, 37–40, 92–93, 30, 96–97. Jackson to James Winchester, November 22, 1814, Miscellaneous Jackson Papers, NYHS.

9. Jackson to Monroe, December 10, 1814; Jackson to Coffee, December 11, 1814; Jackson to Claiborne, September 21, 1814; Jackson, *Correspondence,* II, 111–13, 56; Parton, *Jackson,* II, 62, 31.

10. Jackson to William Carroll, December 16, 1814; Jackson to Claiborne, December 24, 1814; Jackson to Monroe, February 13, 1815; Jackson to Governor Holmes, December 25, 1814; Jackson, *Correspondence,* II, 116, 123–24, 167, 124. Alexander Walker, *Jackson and New Orleans,* p. 151.

11. Parton, *Jackson,* II, 159, 161; Jackson to Monroe, January 2, 1815, Jackson, *Correspondence,* II, 130. *Niles' Weekly Register,* February 11, 1815.

12. Daniel Patterson to Jackson, January 7, 1815, Jackson, *Correspondence,* II, 132. Raleigh, N. C., *The Star,* February 10, 1815. Parton, *Jackson,* II, 197; Jackson to Monroe, January 9, 13, 1815, Jackson, *Correspondence,* II, 136–38, 142–44. Walker, *Jackson and New Orleans,* p. 353. Jackson to Morgan, January 8, 1815, Jackson, *Correspondence,* II, 135.

13. Raleigh, N.C., *The Star,* February 10, 1815; Philadelphia, *Aurora,* February 9, 1815; *Niles' Weekly Register,* February 11, 25, 1815. Parton, Jackson, II, 208–9. Jackson to Monroe, January 19, 1815, Jackson, *Correspondence,* II, 148–49.

14. Parton, *Jackson,* II, 273–74, 246; *Niles Weekly Register,* February 11, February 18, 1815; Monroe to Jackson, February 5, 1815, Jackson, *Correspondence,* II, 158; Epes Sargent, *The Life and Public Services of Henry Clay,* p. 59.

CHAPTER IV

1. Jackson to Lewis, May 8, 1815, Jackson-Lewis Papers, NYPL; Parton, *Jackson* II, 334; *American State Papers, Foreign,* IV, 559; Bassett, *Jackson,* p. 240.

2. *American State Papers, Military Affairs,* I, 687; Monroe to Jackson, December 28, 1817, Monroe Papers, NYPL; Calhoun to Jackson, December 26, 1817; Calhoun to Gaines, December 16, 1817; *American State Papers, Military Affairs,* I, 690, 689. Jackson to Monroe, January 6, 1818; Rhea to Jackson, December 18, 1818; Jackson, *Correspondence,* II, 345, 404.

3. Monroe to Jackson, December 28, 1817, Monroe Papers, NYPL. A different and decidedly anti-Jackson interpretation can be found in Richard R. Stenberg, "Jackson's Rhea Letter Hoax," *Journal of Southern History* (1936). II, 480–96.

4. Joseph Anderson to Jackson, no date, Joint University Library, Nashville. Rhea to Jackson, January 12, 1818; Jackson to Calhoun, April 8, April 20 1818, Jackson, *Correspondence,* II, 348, 358–59, 361–62. James, *Jackson* 288. Parton, *Jackson,* II, 475. Jackson to Calhoun, May 5, 1818, Jackson, *Correspondence,* II, 367; *American State Papers, Military Affairs,* I, 700, 682, 721–34.

5. *American State Papers, Military Affairs,* I, 702; Jose Masot to Jackson, May 22, 24, 1818, Jackson, *Correspondence,* II, 371, 372; Parton, *Jackson,* II, 500; New Orleans *Gazette,* May 12, 1818; New York *Daily Advertiser,* May 28, June 12, 1818; *American State Papers, Foreign,* IV, 495; John Quincy Adams, *Memoirs,* IV, 102–4, 109; Benjamin Butler to Harriet Butler, May 7, 1823, Butler Papers, NYSL; Monroe to Jackson, October 20, 1818, Jackson, *Correspondence,* II, 398; *American State Papers, Foreign,* IV, 545–610; Adams, *Memoirs,* IV, 109.

6. Jackson to Andrew Donelson, January 31, 1819; Jackson to Crawford, June 10, 13, 16, 1816; Crawford to Jackson, July 1, 1816; Jackson, *Correspondence,* II, 408, 243–51. Clay's speech was delivered on January 20, 1819 and can be found in the *Annals of Congress,* 15th Congress, and Session, pp. 631–55. New York *Daily Advertiser,* April 13, 1818. Jackson to Lewis, January 30, 25, 1819, Jackson-Lewis Papers, NYPL.

7. *American State Papers, Foreign,* IV, 616–21; Samuel F. Bemis, *John Quincy Adams,* I, 315; Jackson to James Gadsden, August 1, 1819, Jackson, *Correspondence,* II, 424; Jackson to Lewis, January 30, January 25, 1819, Jackson-Lewis Papers, NYPL; Jackson to Eaton, November 29, 1819, Jackson Papers,

LC; Charles Sellers, "Jackson Men with Feet of Clay," *American Historical Review* (1957), LXII, 537–51.

8. Jackson to Coffee, March 1, 1821, Jackson, *Correspondence*, III, 41; Rachel Jackson to Mrs. Eliza Kingsley, April 27, 1821, quoted in Parton, *Jackson*, II, 595, 641. Jackson to Callava, August 3, 1821; Jackson to Adams, August 26, 1821 and see the June and July correspondence between Jackson and Callava, 1821, Jackson, *Correspondence*, III, 108–11, 112–16, 70–104.

9. On Jackson's tenure in Florida see the sympathetic articles by Herbert J. Doherty in the *Florida Historical Quarterly*, XXXIII, XXXIV, XXXV, 1954–55; but for a critical analysis see *An Examination of Governor Jackson in Florida, passim;* Parton, *Jackson*, II, 639; Bassett, *Jackson*, 317; Jackson to Monroe, October 5, 1821, Miscellaneous Jackson Papers, NYHS.

CHAPTER V

1. John Ward, *Andrew Jackson: Symbol for an Age, passim;* Murray Rothbard, *The Panic of 1819*, pp. 95–96; Martin Van Buren, *Autobiography*, quoted in Bassett, *Jackson*, p. 702; Charles G. Sellers, "Jackson Men With Feet of Clay," *American Historical Review*, (1957), LXII, 537–51. An initial hesitation on Jackson's part to the invitation of the Junto was dissolved when he learned that some of his erstwhile friends were secretly working for Henry Clay in Tennessee. See Jackson to Donelson, August 22, 1822, Jackson, *Correspondence*, III, 174.

2. Jackson to James Gadsden, December 6, 1821, Jackson, *Correspondence*, III, 140–41; George Dangerfield, *Awakening of American Nationalism*, pp. 212 ff.; Robert V. Remini, *Martin Van Buren and the Making of the Democratic Party*, p. 48; Parton, *Jackson*, III, 47–48; *Annals of Congress*, 18th Congress, 1st session, 708, 733.

3. It is probable that Clay did not decide his choice in the House election until his meeting with Adams. Several letters written during the latter part of 1824 repeat his indecision. (See Clay to Hubbard Taylor, December 25; Clay to Peter Porter, December 26, 1824; Clay, *Correspondence*, III, 904, 905.) Only one letter supposedly proves that Clay, as early as December, 1824, "had decided in favor of Mr. Adams in case the contest should be between him and General Jackson;" however, the letter just quoted was actually written by George McClure to Robert S. Rose on September 1, 1827 claiming to quote an earlier Clay

letter dated December 28, 1824. All we have is McClure's word that he is quoting Clay, and since his letter was written in 1827, during the height of the "corrupt bargain" charges, there is sufficient reason to question the authenticity of the 1824 letter. (See McClure to Rose, September 1, 1827, Calvin Colton, *Life and Times of Clay*, I, 387 n.)

4. William Plumer, Jr. to William Plumer, January 24, 1825, Brown, ed., *Missouri Compromises and Presidential Politics*, p. 134; Adams, *Memoirs*, VI, 501; Philip Klein, *President James Buchanan*, p. 52; Jackson to Lewis, December 22, 1823, March 31, 1824, Jackson-Lewis Papers, NYPL; Jackson to Lewis, December 27, 1824, Jackson Papers, ML; Martin Van Buren, *Autobiography*, 152; Shaw Livermore, *Twilight of Federalism*, pp. 180–82.

5. Van Buren, *Autobiography*, p. 152, is the only evidence for this extraordinary story of Van Rensselaer's performance. But it is entirely in keeping with Van Rensselaer's character and personality. Besides, Van Buren had no reason to invent such an outlandish tale. Ben: Perley Poore, *Reminiscences*, p. 26; Jackson to Lewis, February 16, 1825, Miscellaneous Jackson Papers, NYHS; Rufus King to John King, February 27, 1825, King Papers, NYHS; Jackson to Lewis, March 8, 1828, Jackson Papers, ML; Remini, *Election of Jackson*, pp. 51–94.

6. U.S. *Telegraph*, June 16, 1828; Kendall to Blair, February 3, 1829, Blair-Lee Papers, PUL; Jackson to Kendall, September 4, 1827, Jackson-Kendall Papers, LC; *National Journal*, September 4, 1828, May 26, 28, June 2, 16, 1827.

7. U.S. *Telegraph*, April 22, May 26, June 2, 16, 1827; *National Journal*, September 4, 1828; Raleigh, N. C. *Register*, May 15, 1827; Boston *Advertiser*, May 3, 1828; Thurlow Weed, *Autobiography*, p. 297; Henry Stanton, *Random Recollections*, p. 25; Charles McCarthy, "The Antimasonic Party," *American Historical Association, Annual Report* (1902), I, 371–85; Edward Pessen, "The Workingmen's Movement in the Jackson Era," *Mississippi Valley Historical Review* (1956), XLIII, 428–29.

8. Remini, *Election of Jackson*, pp. 114–18, 171–80, 166–95; Parton, *Jackson*, III, 154–64; for a fuller discussion of the tariff of 1828, see Remini, "Martin Van Buren and the Tariff of Abomination," *American Historical Review*, LVII, 1959, 117–29; Thomas S. Hinde to Clay, February 3, 1829, Clay Papers, LC; Jackson to Coffee, January 17, 1829, Jackson, *Correspondence*, IV, 2.

CHAPTER VI

1. E. P. Gaines to Jackson, November 22, 1828, Jackson Papers, LC; Parton, *Jackson*, III, 164–73; Daniel Webster to Mrs. E. Webster, February 19, 1829, Webster, *Correspondence*, I, 470; Ben: Perley Poore, *Reminiscences*, pp. 92–96; Richardson, *Messages and Papers of the Presidents*, II, 1000–1; Margaret B. Smith, *The First Forty Years of Washington Society*, p. 283; James Hamilton, Jr. to Van Buren, March 5, 1829, Van Buren Papers, LC.

2. Charles Wiltse, *John C. Calhoun, Nullifier*, pp. 21–23; Jackson to John C. McLemore, April, 1829, Miscellaneous Jackson Papers, NYHS; Arthur M. Schlesinger, Jr. *The Age of Jackson*, p. 104; Leonard White, *The Jacksonians*, p. 308; Sidney Aronson, *Status and Kinship in the Higher Civil Service*, pp. 158 ff.; Richard P. Longaker, "Was Jackson's Kitchen Cabinet a Cabinet?" *Mississippi Valley Historical Review* (1957), XLIV, 94–104, 107–8.

3. Parton, *Jackson*, III, 185; Cambreleng to Van Buren, January 1, 1829, Van Buren Papers, LC. The ellipsis is Cambreleng's. Donelson to Coffee, September 20, 1829, Donelson Papers, LC; Jackson to Ely, March 23, 1829, quoted in Parton, *Jackson*, III, 187–88, 191, 288; Donelson to Ely, September, 1829, Jackson Papers, LC.

4. Jackson to McLemore, December 25, 1830, Miscellaneous Jackson Papers, NYHS; Narrative of William B. Lewis, October 25, 1859 quoted in Parton, *Jackson*, III, 321, 288, 296; Jackson to McLemore, September 28, 1829, Miscellaneous Jackson Papers, NYHS.

5. Richardson, *Messages and Papers*, II, 1021–22, 1013, 1025.

6. Register of Debates, 21 Congress, 1 Session, pp. 22 ff.; Van Buren, *Autobiography*, 413–14, 416; Van Dusen, *Jacksonian Era*, p. 44.

7. From reading Jackson's correspondence, it is clear that during the Seminole controversy the General was absolutely convinced of Calhoun's support. He probably changed his mind sometime between 1824 and 1828. Jackson to McLemore, December 25, 1830, Miscellaneous Jackson Papers, NYHS. Although it is difficult to believe, Jackson said Calhoun's 52 page reply "surprised, yes astonished me." "Memorandum," 1832, Jackson Papers, LC. See also the Jackson-Calhoun correspondence in May, June, July, and August, 1830, Jackson Papers, LC.

8. Balch to Jackson, September 14, 1831, Jackson Papers, LC.

Washington *Globe*, February 21, 23, March 26, 1831; Benton, *Thirty Years View*, I, 167–80; Van Buren, *Autobiography*, p. 403. For a completely different view of Calhoun's downfall, see Wiltse, *Calhoun, Nullifier*, pp. 26–97.

9. Parton, *Jackson*, III, 345–46; Eaton to Ingham, June 17, 1831 and Ingham to Eaton, June 18, 1831, quoted in Parton, *Jackson*, III, 365, 368; Van Buren, *Autobiography*, pp. 407–8.

CHAPTER VII

1. Felix Grundy to Levi Woodbury, February 7, 1830, Woodbury Papers, LC; Richardson, *Messages and Papers*, II, 1054–55; Van Buren, *Autobiography*, p. 327; Jackson to Kendall, July 23, 1831, Jackson-Kendall Papers, LC; Richardson, *Messages and Papers*, II, 1341. With respect to the Maysville Road Jackson was convinced Clay was attempting to box him in and, as he said, "compel me to approve . . . & thereby acknowledge Mr. Clay's [American System] doctrine or disapprove them upon constitutional grounds." Under the circumstances it is not surprising that Jackson vetoed the bill. Jackson to Overton, May 13, 1830, Overton Papers, TSL.

2. Jackson to Major David Haley, October 15, 1829, Jackson Papers, LC; Cherokee Nation v. Georgia, 5 Peters, 10 ff., and Worcester v. Georgia, 6 Peters, 521 ff. Horace Greeley, *The American Conflict*, I, 106; Grant Foreman, *Indian Removal, passim;* Van Deusen, *Jacksonian Era*, p. 50.

3. Benton, *Thirty Years View*, I, 219; Jackson to Woodbury, September 11, 1832; Jackson to Cass, October 29, 1832; Jackson Papers, LC; Jackson to Donelson, August 30, 1832, Donelson Papers, LC; Chancey Boucher, *Nullification Controversy in South Carolina*, pp. 164–207; Silas Wright to John M. Niles, February 13, 1833, Gideon Welles Papers, LC.

4. Jackson to Poinsett, January 24, 1833, December 9, 1832, Jackson, *Correspendence*, V, 11, IV, 498; Richardson, *Messages and Papers*, II, 1161; Jackson to Coffee, December 14, 1832, Jackson, *Correspondence*, IV, 500; Richardson, *Messages and Papers*, II, 1203–19; Jackson to Edward Livingston, no date, Jackson, *Correspondence*, IV, 494; Francis Lieber to Woodbury, January 13, 1833, Woodbury Papers, LC.

5. Jackson to Poinsett, December 9, 1832, Jackson, *Correspondence* IV, 498; Erastus Smith to John M. Niles, January 4, 1833, Welles Papers, LC; W. B. Lawrence to Albert Gallatin, January 6,

1833, Gallatin Papers, NYHS; Van Deusen, *Life of Henry Clay*, p. 268; Wright to Flagg, January 27, 1833, Flagg Papers NYPL. On the willingness of the administration to compromise the tariff, see Roger Taney's letters to Thomas Ellicott, January 25, 1832, January 22, 1834, Taney Papers, LC; St. Louis, Mo. *Times*, March 2, 30, 1833; Jackson to Nathaniel Macon, August 17, 1833, Jackson Papers, LC; Bemis, *John Quincy Adams*, II, 269.

6. Parton, *Jackson*, III, 488–92, 479; The story of Dr. Jackson's *"sine qua non"* remark was taken from the political humorist, Major Jack Downing. Richardson, *Messages and Papers*, II, 1075, 1200–1, 1153, 1285–87, 1201–2; White, *The Jacksonians*, p. 29.

CHAPTER VIII

1. Richardson, *Messages and Papers*, II, 1304, 1224–38.
2. Leon Schur, "Second Bank of the United States and the Inflation after the War of 1812," *Journal of Political Economy* (1960), LXVIII, pp. 118–34; Ralph Catteral. *The Second Bank of the United States*, pp. 22–92; Walter B. Smith, *Economic Aspects of the Second Bank of the United States*, pp. 134–46; Richardson, *Messages and Papers*, II, 1522, 1327; W. G. Sumner, *A History of Banking in the United States*, p. 194; Taney to Thomas Ellicott, January 25, 1832, Taney Papers, LC; Webster to Biddle, December 21, 1833, Biddle Papers, LC. Jackson to Overton, June 8, 1829, Jacob Dickinson Papers, TSL.
3. Bassett, *Jackson*, 593; Blair to Jackson, August 17, 1830, Blair Papers, LC; Parton, *Jackson*, III, 260–69; Vincent J. Capowski, "The Making of a Jacksonian Democrat: Levi Woodbury," doctoral dissertation, Fordham University 1965, pp. 210–19; Kendall to Jackson, December, 1830 (?) Jackson Papers, LC; Hamilton, *Reminiscences*, p. 69. James A. Hamilton was *ad interim* Secretary of State before Van Buren took over. Taney to Ellicott, December 15, 1831, Taney Papers, LC; Alexander Hamilton to Biddle, December 10, 1829, Edward Everett to Biddle, December 15, 1829, Biddle Papers, LC.
4. Balch to Jackson, January 8, 1830, Jackson, *Correspondence*, IV, 115; "Statement" in Jackson's hand, 1830, Jackson Papers, LC; Govan, *Biddle*, 112, 172; Taney to Ellicott, January 25, 1832, Taney Papers, LC; P. Smith to Biddle, December 7, 1831, N. D. Merth (?) to Biddle, December 14, 1831; Thomas Cadwalader to Biddle, December 21, 1831; Biddle Papers, LC.
5. Clay to Biddle, December 15, 1831, Biddle Papers, LC;

Benton, *Thirty Years View,* I, 235–39; Chambers, *Benton,* p. 181; *Globe,* April 6, May 5, 15, June 3, 8, 1832; Taney to Ellicott, February 20, 1832, Taney Papers, LC; Biddle to Cadwalader, July 3, 1832, Biddle Papers, LC; Van Buren, *Autobiography,* p. 625; J. Herron to Donelson, September 1, 1832, Donelson Papers, LC; John McLean to Biddle, July 29, 1832, Biddle Papers, LC; Silas Wright to John Niles, September 24, 1832, Gideon Welles Papers, LC; Benton, *Thirty Years View,* I, 241–2; Jackson to Kendall, July 23, 1832, Jackson-Kendall Papers, LC.

6. Bassett, *Jackson,* p. 619; Lynn L. Marshall, "The Authorship of Jackson's Bank Veto Message," *Mississippi Valley Historical Review* (1963), L, 466–77; Taney, "Bank War Manuscript," Taney Papers, LC; Richardson, *Messages and Papers,* II, 1140–41, 1145, 1153; Washington *Globe,* July 12, 1832; Biddle to Clay, August 1, 1832, Biddle Papers, LC.

7. An indication of Jackson's participation in the election can be glimpsed from his letter to Kendall, July 23, 1832, Jackson-Kendall Papers, LC; Gammon, *Presidential Election of 1832,* pp. 45–103; Worden Pope to Jackson, August 6, 1831, Jackson Papers, LC. Although Henry Clay is one of the most overrated men in American history, he was a superb politician with good natural instincts for the political game. Thus, it still seems incredible that he could fail to win the presidency—and so many times.

8. Kendall to Gideon Welles, September 12, 30, 1831, October 5, 1832. Welles Papers, LC; Parton, *Jackson,* III, 425–26. On the removal, see the lengthy statements by Taney, McLane, and Woodbury, Jackson Papers, LC.

9. E. Eams to Biddle [1834], Biddle Papers, LC; Frank O. Gatell, "Spoils of the Bank War," *American Historical Review* (1964), LXX, 35 ff.; Biddle to Hopkinson, February 21, 1834, Biddle Papers, LC; Jacob Merriman, "Climax of the Bank War: Biddle's Contraction of 1833–34." *Journal of Political Economy* (1963), LXXI, p. 378–88; Samuel Bell to Joseph Blunt, February 24, 1834, Autograph File, HUL; Bell to Jacob Moore, January 22, 1834. Moore Papers, HUL; Jackson to Andrew Jackson, Jr., February 16, 1834, Jackson, *Correspondence,* V, 249. Harry Scheiber, "Pet Banks in Jacksonian Politics and Finance," *Journal of Economic History* (1963), XXIII, 196 ff.

10. *Register of Debates,* 23rd Congress, 1st Session, 1073–1105; Jackson to William Findlay, August 20, 1834. Jackson Papers, LC; John Van Buren to Martin Van Buren, February 28, 1834,

John Van Buren Papers, Private Collection; Silas M. Stilwell to Biddle, January 27, 1834, Biddle Papers, LC; Butler to Olcott, January 27, 1834, Olcott Papers, CUL; Robert Lytle to Blair, March 5, 1834, Richard M. Johnson to Blair, June, 1835 (?) Blair-Lee Papers, PUL; C. Lawrence to George Newbold, April 1, 1834, Newbold Papers, NYHS; Woodbury to Nathaniel Niles, February 26, 1834, Woodbury Papers, LC; Johnson to Blair, no date, Blair-Lee Papers, PUL.

11. Parton, *Jackson*, III, 549–50; Wise, *Seven Decades*, 106–7; "Recollections of an Old Stager," *Harpers* (1872), XLV, 602; Butler to Olcott, February 1, 1834. Olcott Papers, CUL; L. J. Pease to Welles, January 29, 1834, Welles Papers, LC; J. S. Barbour to James Barbour, January 22, 1834, Barbour Papers, NYPL; Jesse Hoyt to Van Buren, January 29, February 4, 1834, Mackenzie, *The Life and Times of Martin Van Buren*, p. 249; C. Pickings to Woodbury, January 28, 1834, Woodbury Papers, LC; Van Buren, *Autobiography*, 729–30; *Register of Debates*, 23rd Congress, 1st Session, 399–402; H. Binney to Biddle, January 31, 1834, Biddle Papers, LC.

12. Butler to Olcott, March 20, 1834, Olcott Papers, CUL; *Register of Debates*, 23rd Congres, 1st Session, 58–59; Richardson, *Messages and Papers*, II, 1295, 1304–5, 1312; Blair to Jackson, December 14, 1834, Blair Papers, LC; Hopkinson to Biddle, February 11, 1834, Biddle Papers, LC; Jackson to Wolf, February, 1834, Jackson, *Correspondence*, V, 243–44.

13. Taney to Polk, March 11, 1834, Polk Papers, LC; *Register of Debates*, 23rd Congress, 1st Session, 3474–77; Sellers, *Polk*, 221; John Y. Mason to Polk, May 10, 1834, Polk Papers, LC; Rives to Woodbury, May 26, 1834, Jackson to Woodbury, July 3, 1834, Woodbury Papers, LC; Jackson to Kendall, November 24, 1836, Jackson-Kendall Papers, LC; Jackson to Woodbury, July 3, 8, August 15, 1834, Woodbury Papers, LC.

14. Butler to Harriet Butler, January 14, 183 (?), Butler Papers, NYSL; Richardson, *Messages and Papers*, II, 1327; Jackson to Woodbury, August 15, 1834, Woodbury Papers, LC.

CHAPTER IX

1. Alice Felt Tyler, *Freedom's Ferment, passim;* Parton, *Jackson*, III, 487, 616; Washington *Globe*, February 3, 1835; United States *Telegraph*, February 2, 1835.

2. Frank Benns, *The American Struggle for the British West*

Indian Trade, pp. 107–62; William Rives to Van Buren, November 7, December 17, 1829. Dispatches from U. S. Ministers to France, National Archives; Bassett, *Jackson*, p. 667; Richardson, *Messages and Papers*, II, 1319–26; Jackson to Kendall, July 19, 1835, Jackson-Kendall Papers, LC; Richard McLemore, *Franco-American Diplomatic Relations*, pp. 150–209.

3. Parton, *Jackson*, III, 605; "Notes," 1829, Jackson, *Correspondence*, IV, 58–61; Bassett, *Jackson*, p. 675; Butler to Jackson, October 19, 1829, October 7, 1830, August 24, 1831, February 14, 1833, March 7, 1834, Jackson, *Correspondence*, IV, 82, 183, 335–36, V, 17, 251–52; James Callahan, *American Foreign Policy in Mexican Relations*, pp. 62–99.

4. 11 Peters 420 ff.; F. Byrdsall, *History of the Loco-Foco or Equal Rights Party*, pp. 111, 149, 167; Richardson, *Messages and Papers*, II, 1326; Van Deusen, *Jacksonian Era*, pp. 104–7; Sellers, *Polk*, p. 232.

5. Richardson, *Messages and Papers*, II, 1523, 1527; Parton, *Jackson*, III, 627; Benton, *Thirty Years View*, I, 735.

CHAPTER X

1. Jackson to Kendall, March 28, 1837, Jackson-Kendall Papers, LC; Blair to Jackson, May 1, 1837, Blair Papers, LC; Jackson to Blair, April 19, 1841, Jackson, *Correspondence*, VI, 105.

2. Bassett, *Jackson*, p. 746; Jackson to Blair, May 11, 1844, Jackson, *Correspondence*, VI, 286–87; James, *Jackson*, p. 769; Parton, *Jackson*, III, 676.

3. Parton, *Jackson*, III, 671, 677–79.

Bibliography

Those who wish to know something about the life of Andrew Jackson should study two works: James Parton, *Life of Andrew Jackson* (3 vols., New York, 1861) and John S. Bassett, *The Life of Andrew Jackson* (two vols. in one, New York, 1928). If they wish to add some excitement, they should read Marquis James, *Andrew Jackson* (two vols. in one, Indianapolis, 1938). For the history of this period generally, especially to catch the color and splash of the times, there is no better book than Arthur Schlesinger, Jr., *The Age of Jackson* (Boston, 1946). A scholarly and more recent study of the age is Glyndon Van Deusen, *The Jacksonian Era* (New York, 1959) but this book is Whiggish in tone and a bit bland. In some ways the most perceptive book of all about Jackson as President is Leonard White, *The Jacksonians* (New York, 1963).

Lately, historians have been having trouble deciding what Jacksonian Democracy is all about, and one obvious reason for this is their neglect of the central figure of the period. In terms of biography, Jackson himself has not been the subject of serious scholarly study since Bassett first published his work in 1910. None of the biographies subsequently written approach Bassett's scholarship or critical handling of his subject. James' study is an immensely readable book, but despite impressive research he produced a one-dimensional Jackson that is highly partisan. He does not begin to suggest the deep subtleties within Old Hickory. And, as long as the Hero himself remains elusive, Jacksonian Democracy will also be controversial. So disputes about the nature and significance of the Jacksonian movement continues, and readers have a wide range of volumes from which to select interpretative ideas. Among the most interesting and provocative books are: Marvin Meyers, *The Jacksonian Persuasion* (Stanford, 1957); John Ward, *Andrew Jackson, Symbol for an Age* (New York, 1955); Louis Hartz, *The Liberal Tradition in America* (New York, 1955); Frederick Jackson Turner, *Rise of the New West* (New York, 1935); Richard Hofstadter, *The American Political Tradition* (New York, 1948); William MacDonald, *Jacksonian Democracy* (New York, 1907); Bray Hammond, *Banks and*

Politics in America (Princeton, 1957); Vernon L. Parrington, *Main Currents in American Thought*, II (New York, 1954); Carl R. Fish, *The Rise of the Common Man* (New York, 1927); Alice Felt Tyler, *Freedom's Ferment* (Minneapolis, 1944); Algie M. Simons, *Social Forces in American History* (New York, 1911); Louis Hacker, *The Triumph of American Capitalism* (New York, 1940); Frederic A. Ogg, *The Reign of Andrew Jackson* (New York, 1919); Lee Benson, *The Concept of Jacksonian Democracy: New York as a Test Case* (Princeton, 1961); Claude Bowers, *Party Battles of the Jackson Period* (New York, 1922); and H. R. Fraser, *Democracy in the Making* (Indianapolis, 1938).

There are several outstanding biographies of Jackson's contemporaries. Charles M. Wiltse, *John C. Calhoun* (3 vols., Indianapolis, 1944–1951) should be read to get Calhoun's side of the story. For balance, turn to Gerald Capers, *John C. Calhoun: Opportunist* (Gainesville, 1962). Samuel F. Bemis' two volumes on *John Quincy Adams* (New York, 1949, 1956) have no equal and are not likely to be rivaled for some time to come. On Van Buren the best biography is by Edward M. Shepard (Boston, 1899), but it is quite old and rather brief. Robert V. Remini, *Martin Van Buren and the Making of the Democratic Party* (New York, 1959) treats Van Buren's career in the 1820's. A model of judicious biographical writing is Glyndon Van Deusen, *The Life of Henry Clay* (Boston, 1939), but Clement Eaton's, *Henry Clay and the Art of American Politics* (Boston, 1957), and G. R. Poage, *Henry Clay and the Whig Party* (Chapel Hill, 1936) are also valuable. Other excellent biographies include: William N. Chambers, *Old Bullion Benton* (Boston, 1956); Elbert B. Smith, *Magnificent Missourian: The Life of Thomas Hart Benton* (Philadelphia, 1958); Charles G. Sellers, *James K. Polk, Jacksonian* (Princeton, 1957); Eugene I. McCormac, *James K. Polk* (Berkeley, 1922); Philip S. Klein, *President James Buchanan* (University Park, Pa., 1962); Richard N. Current, *Daniel Webster and the Rise of National Conservatism* (Boston, 1955); Claude M. Fuess, *Daniel Webster* (2 vols., Boston, 1930); Carl B. Swisher, *Roger B. Taney* (New York, 1935); Robert Seager, *And Tyler Too* (New York, 1963); Marquis James, *The Raven: A Biography of Sam Houston* (Indianapolis, 1929); William E. Smith, *The Francis Preston Blair Family in Politics* (2 vols., New York, 1933); H. E. Putnam, *Joel R. Poinsett* (Washington, 1935); Oliver P. Chitwood, *John Tyler* (New York, 1939); William B. Hatcher, *Edward Livingston* (Baton

Rouge, 1940); Joseph Parks, *Felix Grundy* (Baton Rouge, 1940); John A. Garraty, *Silas Wright* (New York, 1949); Margaret L. Coit, *John C. Calhoun* (Boston, 1950); and Thomas Govan, *Nicholas Biddle* (Chicago, 1959).

The role of labor in the Jacksonian era has been especially attractive to historians. Walter Hugins, *Jacksonian Democracy and the Working Class* (Stanford, 1960) is a first-rate book. See also Joseph Dorfman, *The Economic Mind in American Civilization*, II (New York, 1946); Philip S. Foner, *History of the Labor Movement in the United States* (New York, 1947); William A. Sullivan, *The Industrial Worker in Pennsylvania* (Harrisburg, 1955); and Joseph G. Rayback, *History of American Labor* (New York, 1959).

Books of special value dealing with certain aspects of the Jacksonian period include: M. Ostrogorski, *Democracy and the Organization of Political Parties* (New York, 1902); Chilton Williamson, *American Suffrage from Property to Democracy* (Princeton, 1960); Shaw Livermore, *The Twilight of Federalism* (Princeton, 1962); George Dangerfield, *The Awakening of American Nationalism* (New York, 1965); Robert V. Remini, *The Election of Andrew Jackson* (Philadelphia, 1963); A. C. Cole, *The Whig Party in the South* (Washington, 1914); George R. Taylor, *The Transportation Revolution* (New York, 1951); R. C. Buley, *The Old Northwest: Pioneer Period* (Indianapolis, 1950); Whitney R. Cross, *The Burned Over District* (Ithaca, 1950); Paul P. Van Riper, *History of the United States Civil Service* (Evanston, 1958); Clement Eaton, *The Growth of Southern Civilization* (New York, 1961); Benjamin A. Hibbard, *A History of Public Land Policies* (New York, 1939); Dwight Dumond, *Anti-Slavery* (Ann Arbor, 1961); Kenneth Stampp, *The Peculiar Institution* (New York, 1956); Henry N. Smith, *Virgin Land* (Cambridge, 1950); Edwin C. McReynolds, *The Seminoles* (Norman, 1957); R. S. Cotterill, *The Southern Indians* (Norman, 1954); Carl R. Fish, *The Civil Service and the Patronage* (New York, 1905); Grant Foreman, *Indian Removal* (Norman, 1932); E. Malcolm Carroll, *Origins of the Whig Party* (Durham, 1925); Russell Nye, *Fettered Freedom* (East Lansing, 1949); and Reginald McGrane, *The Panic of 1837* (Chicago, 1924).

The Bank War has been the subject of intensive debate, much of it anti-Jackson, particularly when written by economic historians. Among the best works are Bray Hammond, already cited;

Walter B. Smith, *Economic Aspects of the Second Bank of the United States* (Cambridge, 1953); Fritz Redlich, *The Moulding of American Banking* (New York, 1947); Ralph C. H. Catterall, *The Second Bank of the United States* (Chicago, 1902); and Sister M. Grace Madeleine, *Monetary and Banking Theories of Jacksonian Democracy* (Philadelphia, 1943).

Of particular value in the study of Jacksonianism are the many monographs dealing with individual states during the era. Arthur W. Thompson's *Jacksonian Democracy on the Florida Frontier* (Gainesville, 1961) is especially perceptive; so, too, are Philip Klein, *Pennsylvania Politics* (Philadelphia, 1940); Charles M. Snyder, *The Jacksonian Heritage: Pennsylvania Politics* (Harrisburg, 1958); Harry R. Stevens, *The Early Jackson Party in Ohio* (Durham, 1957); William S. Hoffman, *Andrew Jackson and North Carolina Politics* (Chapel Hill, 1958); Edwin A. Miles, *Jacksonian Democracy in Mississippi* (Chapel Hill, 1960); Walter R. Fee, *The Transition from Aristocracy to Democracy in New Jersey* (Somerville, N. J., 1933); Thomas P. Abernethy, *From Frontier to Plantation in Tennessee* (Chapel Hill, 1932); Dixon R. Fox, *Decline of Aristocracy in the Politics of New York* (New York, 1965); Henry Mueller, *The Whig Party in Pennsylvania* (New York, 1922); Henry Simms, *Rise of the Whigs in Virginia* (Richmond, 1929); Arthur Darling, *Political Changes in Massachusetts* (New Haven, 1925); Paul Murray, *The Whig Party in Georgia* (Chapel Hill, 1948); and Herbert J. Doherty, *The Whigs of Florida* (Gainesville, 1959).

Several printed collections of sources are available to students, the most valuable of which include: John S. Bassett, ed., *Correspondence of Andrew Jackson* (6 vols., Washington, 1926–1933); John C. Fitzpatrick, ed., *The Autobiography of Martin Van Buren* (Washington, 1920); Clement Eaton, ed., *The Leaven of Democracy* (New York, 1963); Harold C. Syrett, ed., *Andrew Jackson: His Contribution to the American Tradition* (Indianapolis, 1953); and Joseph L. Blau, ed., *Social Theories of Jacksonian Democracy* (New York, 1947). At this writing projects are already under way to publish the important writings of Henry Clay and John C. Calhoun. The great classic of the period is Alexis de Tocqueville, *Democracy In America* (New York, 1954), a book with which all students can begin and later close their study of the Age of Jackson.

Index